T0275619

GUT MICROBIOTA

GUT MICROBIOTA

Interactive Effects on Nutrition and Health

EDWARD ISHIGURO

NATASHA HASKEY

KRISTINA CAMPBELL

Academic Press is an imprint of Elsevier
125 London Wall, London EC2Y 5AS, United Kingdom
525 B Street, Suite 1800, San Diego, CA 92101-4495, United States
50 Hampshire Street, 5th Floor, Cambridge, MA 02139, United States
The Boulevard, Langford Lane, Kidlington, Oxford OX5 1GB, United Kingdom

Notices

Knowledge and best practice in this field are constantly changing. As new research and experience broaden our understanding, changes in research methods, professional practices, or medical treatment may become necessary.

Practitioners and researchers must always rely on their own experience and knowledge in evaluating and using any information, methods, compounds, or experiments described herein. In using such information or methods they should be mindful of their own safety and the safety of others, including parties for whom they have a professional responsibility.

To the fullest extent of the law, neither the Publisher nor the authors, contributors, or editors, assume any liability for any injury and/or damage to persons or property as a matter of products liability, negligence or otherwise, or from any use or operation of any methods, products, instructions, or ideas contained in the material herein.

Library of Congress Cataloging-in-Publication Data
A catalog record for this book is available from the Library of Congress

British Library Cataloguing-in-Publication Data
A catalogue record for this book is available from the British Library

ISBN: 978-0-12-810541-2

Publisher: Andre Gerhard Wolff
Acquisition Editor: Megan Ball
Editorial Project Manager: Jaclyn Truesdell
Production Project Manager: Anusha Sambamoorthy
Cover Designer: Matthew Limbert

Typeset by SPi Global, India

DEDICATION

To Ann, for many years of unequivocal support.

- Edward Ishiguro

I would like to thank my husband, Ryan Haskey, who provided encouragement and support to write this book. I would also like to express my gratitude to all the dietitians who share a love for the nutrition profession and have motivated me to pursue my passion for gut health. And finally my mother, who always believes in all my endeavors.

- Natasha Haskey

To Rob and Connie, who taught me to work, and whose love and support I never doubt.

- Kristina Campbell

CONTENTS

PREFACE

The existence of dense populations of microorganisms in the digestive tracts of animals was established over a century ago. In humans, it has also been long known that the microbial colonization of the gut is an event that takes place starting from birth. But little progress was made in understanding the significance of these microbes until the development of the first stable populations of germ-free rodents in the 1940s. Over the ensuing decades, germ-free animal research revealed striking abnormalities in immune function, physiological activity, and anatomical development—properties not exhibited by animals with a microbiota. Moreover, these abnormalities were at least partially reversed by the intentional introduction of a gut microbial population, indicating that gut microbes play critical roles in a variety of important functions. The complexity of the gut microbiota prevented further mechanistic understanding of their roles until key molecular biological techniques were developed in the latter part of the 20th century; then, the Human Genome Project in 2008 kicked off a period of remarkable advancement.

No longer is the gastrointestinal tract seen as just a "tube" where nutrients are digested and absorbed. The complexity of its function is evident now that we understand that substances passing through the gastrointestinal tract have the potential to interact with our microbiota and have profound influences on health. Intestinal microbiology is now one of the most important areas of medical research—with nutritional insights at the forefront of scientific discovery, thanks to findings on the gut microbiome and diet.

A wealth of data is emerging on the gut microbiome—thousands of scientific articles are being published each year, making it incredibly challenging to stay up to date. Health professionals and scientists are being bombarded with questions from the media and the general public, but knowledge translation is lagging behind scientific discovery. And while some are pushing back on the microbiome "hype" seen in some popular books and media, few comprehensive resources exist that synthesize the evidence-based information in a clinically useful way. Our mission in writing this book is to bridge the gap between scientific work and knowledge translation. This book targets the health professional, scientist, or student wanting a scientific touchstone on nutrition and the intestinal microbiome.

KEY FEATURES OF THE BOOK

This first edition of *Gut Microbiota: Interactive Effects on Nutrition and Health* focuses on the fascinating intestinal microbiome as it relates to nutrition. It covers the core science in the microbiome field and draws links between the microbiome and nutrition in medicine. Our goal is to reflect the most current state of evidence available in the field and summarize it in a concise manner. The early chapters introduce key concepts about the microbiome, and the later chapters focus on the application of the gut microbiome and nutrition science. Key objectives are emphasized at the beginning of each chapter, while sidebars highlight and provide more detail about important concepts. Both human studies and animal studies (where appropriate) are discussed throughout the work.

OVERVIEW OF THE CHAPTERS

Chapters 1–5 cover general concepts about the microbiome, laying the foundation for the diet chapters that follow. Chapter 1 begins by discussing important definitions and concepts relating to the microbiome, with an explanation of why and how microbiome research has proceeded so rapidly over the past decade. Chapter 2 covers the gut microbiota specifically: microbiota composition and functionality in the gastrointestinal tract, including the role of the microbiota in immunity. In Chapter 3, readers will follow the acquisition and age-related changes of the gut microbiota throughout life, from infancy to older adulthood. Chapter 4 describes how the gut microbiota is linked with different states of health and disease. Chapter 5 explains the environmental factors that appear to influence the gut microbiota, including antibiotics and other drugs. Chapter 6 delves into nutrition and discusses how various food components impact microbiota composition and function in humans and what kinds of diets may disrupt the normal microbiota.

The remaining four chapters focus on the applications of gut microbiota and nutrition science. Chapter 7 describes how the gut microbiota in disease can be manipulated through interventions, such as probiotics. Chapter 8 covers the most commonly asked questions about nutrition and the microbiome and strives to provide practical, evidence-based diet recommendations. Chapter 9 addresses how gut microbiota research applies to food science—for example, in the development of functional foods. Chapter 10 wraps up the book with a glimpse into the future of gut microbiota and nutrition, an attempt to identify gaps in the research on gut microbiota and nutrition, and a discussion of the major questions to be answered in the coming years.

As we continue to track the ideas and progress in this field, we value readers' input and feedback. Feel free to contact us!

Edward Ishiguro
Natasha Haskey
Kristina Campbell

ACKNOWLEDGMENTS

We appreciate all the colleagues who provided words of wisdom, encouragement, and support during the writing of this book. Thank you for selecting it—we hope it will serve as a useful resource for your daily practice.

We would like to extend a special thanks to David Despain for providing valuable context for Chapter 9.

CHAPTER 1

An Overview of the Human Microbiome

Objectives

- To understand the relationship between the human body and its associated microorganisms.
- To become familiar with the terminology of the human microbiome and with the methods that enable it to be studied.
- To learn about large-scale projects aimed at characterizing the "normal" human microbiome.

WHAT IS A HUMAN?

The human species, *Homo sapiens*, is usually defined as a large-brained bipedal primate with a capacity for language and a knack for using complex tools. A human's 22,000 genes account for hair and eye color, predisposition to disease, cognitive ability, and even aspects of personality. Yet recent discoveries indicate that this description of a human is incomplete.

Humans are covered, inside and out, with a living layer of microbes: bacteria, archaea, fungi, and viruses. Although these microbes are too small to be seen with the naked eye, they are a fundamental part of our human biology. No human or human ancestor has lived without this collection of microbes (Moeller et al., 2016); it has evolved with us over millions of years and is thought to be as necessary for health and survival as a major organ system. These microbes live in an ecosystem with the human at its core; the human is the **host**, providing the resources the microbes need to sustain themselves.

MICROBIOLOGICAL METHODS

Setting the Stage for Discovery of the Human Microbiome

Antonie van Leeuwenhoek—a cloth merchant by trade—is credited for the discovery of single-celled microorganisms, which he called "wee animalcules" (little animals) (Dobell, 1932). With a simple, personally handcrafted

Gut Microbiota
https://doi.org/10.1016/B978-0-12-810541-2.00001-4

microscope, in the late 1600s, he documented the presence of microorganisms in samples collected from a variety of sources. Leeuwenhoek was the first to observe microorganisms in the human body; he found them in dental plaque and in a stool sample on one occasion when he was ill with diarrhea. About two centuries would pass before techniques were developed to explore the significance of Leeuwenhoek's observations.

Robert Koch's extraordinary research career spanned the greater part of an era dubbed the "golden age of bacteriology," 1876–1906 (Blevins et al., 2010). In 1876, Koch published a paper demonstrating that anthrax was caused by the bacterium *Bacillus anthracis*, providing the first proof for the germ theory of disease (Blevins et al., 2010). But his original methods for laboratory cultivation of bacteria were crude and inadequate for routine use, hindering his further progress. To obtain pure cultures—that is, cultures composed of a single bacterial species—he required a solid medium that would support bacterial growth. His attempts to grow bacteria on the surface of slices of potato or on media solidified with gelatin were unsuccessful. The breakthrough occurred when Fannie Angelina Hesse, the wife of Koch's associate Walther Hesse, suggested the use of agar to solidify liquid bacteriologic media (Hesse & Gröschel, 1992). Armed with this new medium, Koch and his colleagues developed methods for isolating and studying pure cultures of bacteria. The impact on medical microbiology was immediate, and between 1878 and 1906, nineteen new bacterial pathogens were linked to specific infectious diseases. These techniques, augmented and supplemented with advances in microscopy and microbial biochemistry, endure in modern microbiology laboratories. They not only have formed the basis for the culture-dependent microbiology but also have fostered the expansion of microbiology beyond pathogenicity into diverse fields like biochemistry, genetics, ecology, and biotechnology.

By the 1980s, however, the growing awareness of the great abundance, diversity, and environmental ubiquity of microorganisms (Whitman et al., 1998) prompted a shift in research strategy. The complexity of microbial communities in their natural habitats was exemplified by the observation that most of the microscopically observable microorganisms in an environmental sample could not be cultured in the laboratory. This discrepancy between microbes that could be observed and those that could be cultured was a phenomenon termed the "great plate count anomaly" (Staley & Konopka, 1985). Usually, between 1.0% and 0.1% of the total bacteria could be accounted for by the standard plating method. Thus, scientists realized that culture-dependent methods alone would be completely inadequate for studying complex populations such as those populating the human body. This prompted a search for alternative methods.

Culture-Independent Microbiology for Exploring the Human Microbiome

Several significant discoveries paved the way for the development of culture-independent methods, which, for the first time, allowed access to the unculturable fraction of natural microbial populations like the human microbiota. The most significant early contribution was made by Carl Woese (Pace et al., 2012). In the 1960s, Woese began studying the evolution of microorganisms—asking seemingly intractable questions that could not be answered by classic paleontology methods. Microbes, after all, not only were unicellular and microscopic but also were soft-bodied and left no fossil record except in a few extremely rare instances. Even if they were successfully fossilized, they would hardly ever display unique recognizable morphological characteristics distinctive enough to permit species identification. Woese consequently used a molecular phylogenetic approach for tracing evolutionary history. In this approach to tracking microorganisms' evolution, he took cellular ribosomes (the most abundant organelles in all forms of cellular life, performing the essential function of protein biosynthesis) and undertook a comparative analysis of the sequences of one component: the small subunit ribosomal RNAs or SSU rRNAs. Woese reasoned that the similarities and differences between these sequences (i.e., the order of the four chemical bases—adenine, uracil (or thymine in DNA), cytosine, and guanine) would reflect the phylogenetic relationships of the organisms from which they were obtained.

Over many years, Woese and his associates collected and comparatively analyzed the sequences of SSU rRNAs from numerous species of microorganisms. SSU rRNA turned out to be, in Woese's own words, "the ultimate molecular chronometer" (Woese, 1987). There are two forms of SSU rRNAs, designated 18S and 16S. Eukaryotic cells, characterized by genomes enclosed within nuclear membranes, have 18S rRNA, and the morphologically simpler prokaryotic cells that lack nuclear envelopes have 16S rRNA. From analyses of 16S rRNA sequences, Woese and his coworkers discovered that there were actually two distinct groups of prokaryotic cells: the bacteria (originally named eubacteria) and a newly recognized group that was named the archaebacteria (Woese & Fox, 1977). In 1990, the group proposed a new taxonomic scheme to cover all forms of life on Earth, composed of three domains—the domain Eucaryota that included all eukaryotic cells and the two prokaryotic domains, the Bacteria and the Archaea (Woese et al., 1990). The SSU rRNA sequences not only contained unique short sequences that defined the three domains but also contained unique sequences that permitted assignment of cells to specific phyla (Woese, 1987). A universal phylogenetic tree based on SSU rRNA sequences is shown in Fig. 1.1.

The original pioneering studies by Woese involved laborious direct sequencing of SSU rRNA purified from ribosomes. Several key developments expanded the range of applications for SSU rRNA analysis (Escobar-Zepeda et al., 2015). The inventions of DNA sequencing by Sanger in 1977 and of polymerase chain reaction (a method permitting the amplification of small amounts of any desired DNA sequence by several orders of magnitude) by Mullis in 1980 permitted the cloning and sequencing of SSU rRNA genes from DNA samples extracted directly from environmental samples. For the first time, this procedure allowed for the characterization of complex microbial communities without need for prior microbial cultivation. The 21st century brought on rapid technological advances such as high-throughput next-generation DNA sequencing and enhanced computational methods to interpret the DNA sequence information. These

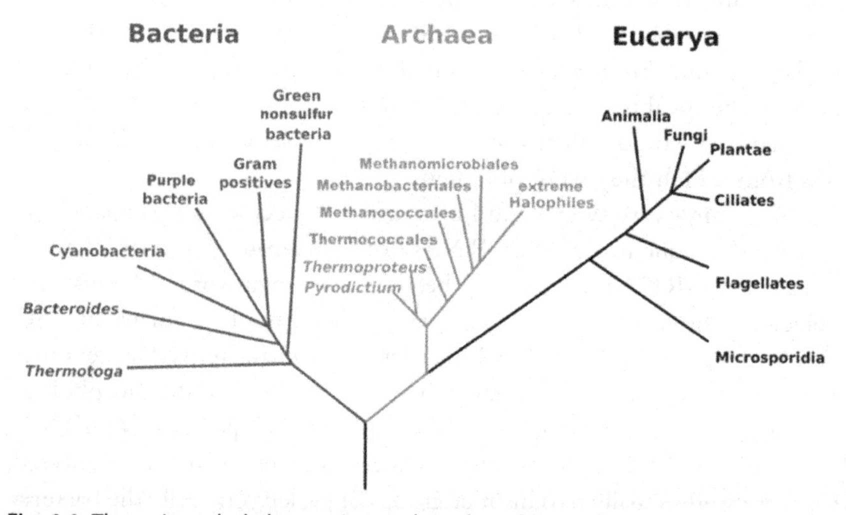

Fig. 1.1 The universal phylogenetic tree based on SSU rRNA sequence analyses accounts for all life forms on Earth. At the root of the tree is the hypothetical last universal common ancestor, and each branch is represented by different phylogenetic groups. The lengths of the branches reflect the amount of evolutionary time separating them. *(From Maulucioni (CC BY-SA 3.0).)*

methods significantly expanded the analysis of microbial community DNA extracted directly from environmental samples beyond SSU rRNA genes; it enabled the sequencing of entire genomes, a procedure termed **metagenomics** (Handelsman, 2004). In **whole-genome shotgun sequencing**, DNA sequences are randomly broken up (shotgunned) into smaller DNA fragments; computer programs reassemble the complete sequence by taking these fragments and looking for regions of overlap. Metagenome sequences provide information about which bacteria exist in a microbial population and at least a partial prediction of what functions are encoded by its genes. These new methods led to an era in which microbiologists had broken away from culture dependence and could now "see" the vast majority of nonculturable microbes that constituted complex microbial communities like the human gut microbiome, and the genetic potentials of the individual community members (Qin et al., 2010). Of particular interest are the metabolic capabilities of the microbiome, their interaction with human metabolism, and their influence on human health.

The Return to Culture Dependence

Microbiome research, ironically, has returned to a phase where culturability will be desirable or even necessary: for example, to determine the individual phenotypes of the many species that constitute the gut microbiota. The murine gut species segmented filamentous bacteria (SFB), for example, which have unique interactions with the immune system (stimulating maturation of B and T cells and increasing small intestinal Th17 responses), were finally cultured after more than 50 years of attempts (Schnupf et al., 2015; Ericsson et al., 2015). A recent report indicated that, in contrast with common belief, the majority of bacteria in fresh fecal samples can indeed be cultured. Researchers cultured these bacteria on a single medium following a relatively simple procedure (Browne et al., 2016). Interestingly, over half of the bacteria isolated were capable of forming resistant spores. The researchers demonstrated that this property significantly promoted survival of the bacteria outside of the gut environment, and they suggested that this may play a role in person-to-person dissemination of these species.

TERMINOLOGY

Four basic categories of microorganisms live in and on the human body: bacteria, archaea, eukaryotes (which include fungi), and viruses. The words "microbes" and "microorganisms" are used interchangeably to refer to all four categories.

The term "microflora" is often used as if synonymous with "microbiota."The original definition of microflora dates back to the early 1600s and originates from the Latin word "flor", meaning flower. Although the definition has evolved, some dictionaries still refer to microflora as "microscopic plants or the plants or flora of a microhabitat." These definitions and their origins make it obvious that microflora refers to plants and not microbes (Marchesi & Ravel, 2015); the assemblage of microbes living in a habitat is now referred to as **microbiota** (either singular or plural).

The word microbiome, while frequently used, does not merely refer to the microbes themselves. **Microbiome** has one of two meanings, depending on context. The first meaning is an inclusive one that refers to an entire habitat where microorganisms dwell, and encompasses the microbes, their genomes, and the surrounding environment. The second, narrower, meaning is "the collection of genes and genomes of members of a microbiota."

Microbiota **composition** is the list of microbes living in a particular habitat. Identifying, naming, and classifying microorganisms—an activity called **taxonomy**—are important foundations on which scientists base their observations. Microbes can be named at different taxonomic levels, from broader to more specific: domains, phyla, classes, orders, families, genera, and species and strains (see Table 1.1 for an example of a bacterium and the taxonomic levels at which it can be named). Researchers generally gain richer information when they identify more specific taxonomic categories—that is, when their data have greater resolution.

Microbiota **function**, on the other hand, is a list of what the microbes in an environment can do. To discover microbiota function, metagenomic approaches (as described above) are used to extract and clone the DNA from an assemblage of microorganisms to study the genomes and genes of

Table 1.1 Example illustrating the names of a bacterium (*Escherichia coli* K-12) at different taxonomic levels

Taxonomic level	Name
Domain	Bacteria
Phylum	Proteobacteria
Class	Gammaproteobacteria
Order	Enterobacteriales
Family	Enterobacteriaceae
Genus	*Escherichia*
Species	*Escherichia coli*
Strain	*Escherichia coli* K-12

its members; these allow researchers to create catalogs of what bacteria can do based on the genes they have (Marchesi et al., 2016).

The microbes in and on an average adult human body comprise 1%–3% of body mass (National Institutes of Health, 2012), with the bacteria vastly outnumbering the other kinds of microorganisms. Recent estimates put the ratio of bacterial cells to human cells at around 1.3 to 1 (Sender et al., 2016)—not 10 to 1, which was a ubiquitous but likely inaccurate estimate.

CHARACTERIZATION OF THE "NORMAL" HUMAN MICROBIOME

A first step in the exploration of the human microbiome was an attempt to characterize "normal," including the range of variation that can be present in different human populations. Although this enormous task is still far from complete today, major advances in understanding the microbiome occurred as a result of two large-scale projects: the Human Microbiome Project (HMP) (Turnbaugh et al., 2007; Methé et al., 2012; Huttenhower et al., 2012) and the European Metagenomics of the Human Intestinal Tract (MetaHIT) project (Qin et al., 2010). The researchers involved in these two projects collected samples from the gut and other body sites of healthy individuals—a total of 2000 people spanning multiple continents (Lloyd-Price et al., 2016).

Human Microbiome Project

The US National Institutes of Health (NIH) launched the HMP near the end of 2007 (NIH HMP Working Group et al., 2009). The goal of the 5-year, $150 million project was to generate resources that would enable the comprehensive characterization of the normal human microbiome and the analysis of its role in human health and disease.

To define the healthy human microbiome, HMP researchers sampled 242 volunteers (129 male and 113 female) from two distinct geographic locations in the United States: Baylor College of Medicine and Washington University School of Medicine (NIH HMP Working Group et al., 2009). Volunteers were screened for the absence of disease, and as such, they were classified as "healthy." Researchers collected over 11,000 samples, with up to three samples from each volunteer at sites such as the mouth, nose, skin (two behind each ear and each inner elbow), and lower intestine (stool), in addition to three vaginal sites in women. In order to evaluate within-subject stability of the microbiome, a subset of individuals (N = 131) were sampled at an additional time point (mean 219 (sd. 69) days after first sampling).

To characterize the microbiota, the method of analysis was 16S rRNA gene analysis, with a subset of the samples shotgun sequenced for metagenomics (Turnbaugh et al., 2007).

In 2012, the HMP consortium reported that they had defined the microbial taxa and genes at each body habitat (Huttenhower et al., 2012). HMP was the first large-scale study to catalog the bacterial taxa associated with different body habitats on the healthy human.

The researchers established that there was substantial variation in microbial community composition and diversity at different body habitats. For example, oral and intestinal communities were especially diverse in terms of community membership, whereas vaginal communities harbored very simple communities. No taxa were universally present in all body habitats and individuals. Each habitat, tested at a single time point, was characterized by one or a few similar taxa; however, diversity and abundance of each habitat's "signature" microbes varied widely even among healthy subjects, with strong niche specialization observed both within an individual and from individual to individual. For those sites tested at a second time point, community variation within an individual was consistently lower than community variation between individuals, suggesting that the uniqueness of one's microbial community is stable over time. The authors concluded this stability could be a feature of the human microbiome specifically associated with health.

HMP was also one of the first large-scale studies to include both marker gene and metagenomic data across body habitats (Huttenhower et al., 2012). In contrast with the variation in relative abundance of bacterial taxa across body habitats and samples, the variation in relative abundance of microbial genes was much smaller among individuals. This finding suggests the existence of functional redundancy—different metabolically active bacteria performing similar functions in different individuals. The prevalence of low-abundance genes varied the most from habitat to habitat, so researchers speculated that the functions of these genes correspond to body-niche-specific activities.

The HMP—a preliminary characterization of the microbiota of healthy adults in a Western population—was an important basis for understanding not only the relationships among microbes but also the relationship between the healthy microbiome and clinical parameters. The observed individual variation in bacterial taxa is critical to take into account when advancing knowledge about microbiome-based disorders. Ultimately, the

HMP created an extensive catalog of taxa, pathways, and genes for use as a reference in subsequent research studies.

Metagenomics of the Human Intestinal Tract

In early 2008, the European Commission and China initiated the project called Metagenomics of the Human Intestinal Tract (MetaHIT), which had both a narrower and a broader goal. Rather than focusing on the microbiome of different body sites, MetaHIT focused on the gut and aimed to link the genes of the human intestinal microbiota to a broad range of states of both health and disease. The research group consisted of representatives from eight countries, linking with 14 partners from academia and industry. Approximately 22 million dollars was dedicated to this project and it was financed mainly by the European Union under the Seventh Framework Program for Research and Technological Development (FP7).

The goal of MetaHIT was to establish associations between the genes of the human intestinal microbiota and health and disease (Qin et al., 2010). To achieve this, fecal specimens were collected from 124 healthy, overweight, and obese adults, and those with inflammatory bowel disease (IBD), from Denmark and Spain. The focus on IBD and obesity was due to the societal impact of these diseases in Europe. MetaHIT categorized a total of 3.3 million nonredundant genes, representing almost 200 times the quantity of microbial DNA sequences reported in all previous studies (Robles-Alonso & Guarner, 2014). Remarkably, it was revealed that this gene set was 150 times larger than the human gene complement. The MetaHIT consortium also discovered that each individual carried over 536,000 prevalent unique genes, indicating that most of the gene pool of 3.3 million was shared by the entire cohort of humans. Almost 40% of the genes from each individual were shared with at least half of the individuals of the cohort.

Most of the genes cataloged were of bacterial origin (99.1%); next to these, the rest were mostly archaeal, with only 0.1% of microorganisms being of eukaryotic and viral origins. The entire cohort harbored ~1000 prevalent bacterial species, with each individual harboring at least 160 species in their gastrointestinal tract. Bacteria from the Bacteroidetes and Firmicutes phyla were found in highest abundance, but the relative proportion varied among individuals. The MetaHIT project yielded the first catalog of reference genes for the human gut microbiome—3.3 million nonredundant genes (Ursell et al., 2012).

Continuation of Microbiome Discovery

In 2012, a large Chinese study carried out a metagenome-wide association study using 368 stool samples of Chinese individuals with type 2 diabetes (T2D) and nondiabetic controls (Qin et al., 2012). Genetic and functional components of the gut metagenome associated with T2D were identified. This study was important as it provided an updated human microbial gene reference set, adding information from both a new ethnicity and from T2D patients to the existing catalog developed by HMP (Huttenhower et al., 2012) and MetaHIT (Qin et al., 2010). An additional 145 gut metagenomes were added to the nonredundant gene catalog as a result of this study.

Between 2008 and 2013, the ELDERMET project based in Ireland recruited almost 500 elderly subjects, aged 65 years and older, across a range of health states from the very frail to the very fit, half of whom were studied at multiple time points. ELDERMET provided important answers to questions regarding the composition and function of the normal microbiota in elderly individuals (Claesson et al., 2011), showing, for example, a relationship between the composition of the intestinal microbiota, diet, and health in this older population (Claesson et al., 2012).

The integrated gene catalog (IGC) contains the most comprehensive intercontinental catalog of reference genes for the gut microbiota discovered to date (Costalonga & Herzberg, 2014; Li et al., 2014). IGC includes a combined set of metagonomic sequencing data from 1267 gut metagenomes from 1070 individuals, including European samples from MetaHIT, American samples from HMP, and samples from a large diabetes study in China, to create a nonredundant gene catalog of 9.8 million microbial genes. Each sample contained about 750,000 genes or about 30 times the number of genes in the human genome, and <300,000 genes were shared by >50% of individuals. The majority of the new genes identified in this latest study were relatively rare, found in <1% of individuals. The IGC also demonstrated through analyses of samples from Danish and Chinese individuals that population-specific characteristics of gut microbiota exist.

In 2010 the Human Oral Microbiome Database (Dewhirst et al., 2010) was launched. It is a web-accessible collection of information on the ~700 prokaryote species detected in the human oral cavity. This curated description of the oral microbiome was created from 16S rRNA gene sequence data and aimed to determine the relative abundance of taxa in the oral cavity and identify new candidate taxa.

Data now suggest that cataloging of the human gut microbiome is entering the stage for identification of rare or individual-specific genes instead of common and shared genes. Characterization of the microbiome in healthy individuals and those living with chronic conditions is an important initial step in understanding the role of the microbiome in contributing to health and disease (Shreiner et al., 2015). Table 1.2 outlines some of the important programs around the world currently undertaking large-scale human microbiome research. Results from many of these projects are pending.

OTUs, Richness, Evenness, and Diversity

When researchers sample a community, they take bacterial sequence data and cluster together members that probably belong to the same taxonomic group: in other words, they define OTUs—**operational taxonomic units**—based on a predetermined similarity threshold (e.g., 97% similarity). The OTUs are similar bacterial individuals, which could be phyla or species, for example. In a sample, OTUs may range from rare to abundant.

Richness is the number of species in a biological community, not taking into account the abundance of each one, while **evenness** is the number of individuals from each species in the community—the distribution of OTU abundance.

Diversity of a bacterial sample is usually represented by the Shannon index (H′, also called the Shannon-Wiener or Shannon-Weaver index) (Hill et al., 2003). Taking into account the number of species and how individual bacteria are distributed among those species, the index measures how statistically difficult it would be to predict the identity of the next bacterial individual sampled, given what is already known about the community. The more rare species there are in the sample, the higher the value. The Shannon index positively correlates with both richness and evenness.

Some scientists caution against reliance on a single number to represent the complex relationships and interactions in a microbial ecological community, and thus, more sophisticated methods, such as species abundance models (which show how abundance is distributed in a population), can also be used to represent diversity (Hill et al., 2003).

Table 1.2 Current programs undertaking large-scale human microbiome research

Program name	Countries involved	Focus	Website
International Human Microbiome Consortium (IHMC)	Australia, Canada, China, France, Gambia, Germany, Kazakhstan, Ireland, Japan, South Korea, Spain, United States	IHMC's efforts are focused on generating a shared comprehensive data resource that will enable investigators to characterize the relationship between the composition of the human microbiome and human health and disease (from 2007 to present)	http://www.human-microbiome.org/
NIH HMP—Project 2	United States	HMP-2 aims to characterize microbial communities found at multiple human body sites and to look for correlations between changes in the microbiome and human health (from 2013 to present)	http://ihmpdcc.org/
EC—"MyNewGut" program	Austria, Australia, Belgium, Canada, Denmark, France, Germany, Ireland, Italy, The Netherlands, New Zealand, Serbia, Spain, United Kingdom, United States	This consortium is undertaking research on the gut microbiome's role in energy balance and brain development/function and development of diet-related diseases and behavior (from 2013 to present)	http://cordis.europa.eu/project/rcn/111044_en.html
APC Microbiome Institute	Ireland	APC is exploring the role of gastrointestinal bacteria (the microbiome) in health and disease (from 2013 to present)	http://www.sfi.ie/assets/media/files/downloads/Investments/APC.pdf

MetaGenoPolis program (MGP)	France	This project aims to establish the impact of the human gut microbiota on health and disease, applying quantitative and functional metagenomics technologies (from 2013 to present)	http://mgps.eu/index.php?id=accueil
Canadian Microbiome Initiative	Canada (with international collaboration)	CMI analyzes and characterizes the microorganisms that colonize the human body and their potential alteration during chronic disease states (from 2008 to present)	http://www.cihr-irsc.gc.ca/e/39939.html
EC—Joint Action "Intestinal Microbiomics" program	International collaboration	This program promotes multidisciplinary transnational research to share and integrate data for performing meta-analyses, and standardizes methods to analyze and understand the human diet-gut microbiome interaction (from 2016 to present)	https://www.healthydietforhealthylife.eu/index.php/joint-actions/microbiomics

Adapted from Stulberg, E., et al., 2016. An assessment of US microbiome research. Nat. Microbiol. 1(1), 15015. Available from: http://www.nature.com/articles/nmicrobiol201515.

Hologenome Model of Evolution

New data on how bacteria interact with human genes have forced scientists to refine their understanding of how humans evolved as a species: the microbiome is now considered a key part of animal and human evolution, with resident microbes facilitating the behavioral and physical adaptation of a host to its environment at different points in time. As shown in Fig. 1.2, the host along with all its associated microorganisms is known as the **holobiont**; the **hologenome model** considers the host genome and microbiome together as a unit of evolution that undergoes selection (Zilber-Rosenberg & Rosenberg, 2008).

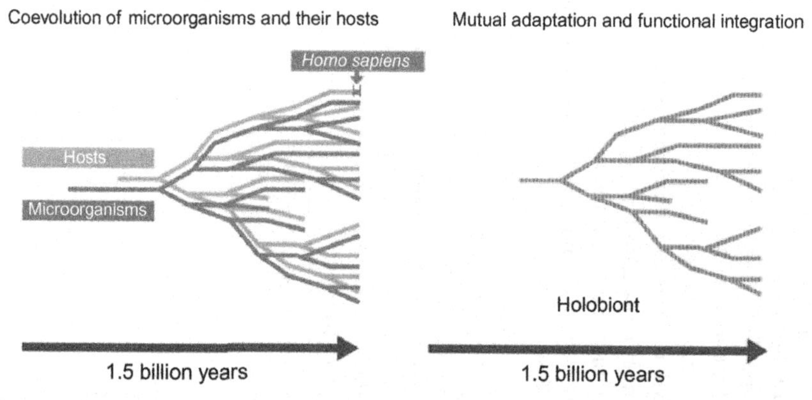

Fig. 1.2 Relationship between microorganisms and hosts through time. *(From Kilian, M., Chapple, I.L.C., Hannig, M., Marsh, P.D., Meuric, V., Pedersen, A.M.L., Tonetti, M.S., Wade, W.G., Zaura, E., 2017. The oral microbiome—an update for oral healthcare professionals. Br. Dent. J. 221, 657–666.)*

Sequencing Scrutiny

An active topic of debate among scientists is the circumstances under which it is advisable to use 16S rRNA (16S) sequencing versus whole-genome shotgun (WGS) sequencing for microbiota surveys. 16S sequencing is currently used in the majority of microbiota studies because of its time and cost-efficiency, even though it yields only crude taxonomic classifications and metagenomics inferences (Janda & Abbott, 2007). Meanwhile, WGS yields reliable strain-level information and data about microbial function but requires greater computational expertise. WGS may be used more often if the cost decreases and when advanced bioinformatics tools are readily available to more researchers.

REFERENCES

Blevins, S.M., et al., 2010. Robert Koch and the "golden age" of bacteriology. Int. J. Infect. Dis. 14 (9), e744–e751. Available from: http://www.ncbi.nlm.nih.gov/pubmed/20413340.

Browne, H.P., et al., 2016. Culturing of "unculturable" human microbiota reveals novel taxa and extensive sporulation. Nature 533 (7604), 543–546. Available from: http://www.nature.com/doifinder/10.1038/nature17645.

Claesson, M.J., et al., 2011. Composition, variability, and temporal stability of the intestinal microbiota of the elderly. Proc. Natl. Acad. Sci. U. S. A. (Suppl. 1), 4586–4591. Available from: http://www.ncbi.nlm.nih.gov/pubmed/20571116.

Claesson, M.J., et al., 2012. Gut microbiota composition correlates with diet and health in the elderly. Nature 488 (7410), 178–184.

Costalonga, M., Herzberg, M.C., 2014. The oral microbiome and the immunobiology of periodontal disease and caries. Immunol. Lett. 162 (2 Pt A), 22–38. Available from: http://www.ncbi.nlm.nih.gov/pubmed/25447398.

Dewhirst, F.E., et al., 2010. The human oral microbiome. J. Bacteriol. 192 (19), 5002–5017. Available from: http://www.ncbi.nlm.nih.gov/pubmed/20656903.

Dobell, C., 1932. Antony van Leeuwenhoek and His "Little Animals": Being Some Account of the Father of Protozoology and Bacteriology and His Multifarious Discoveries in These Disciplines. Harcourt, Brace and Company, New York. Available from: https://archive.org/details/antonyvanleeuwen00dobe.

Ericsson, et al., 2015. Isolation of segmented filamentous bacteria from complex gut microbiota. BioTechniques 59 (2), 94–98. Available from: http://www.biotechniques.com/BiotechniquesJournal/2015/August/Isolation-of-segmented-filamentous-bacteria-from-complex-gut-microbiota/biotechniques-359879.html.

Escobar-Zepeda, A., Vera-Ponce de León, A., Sanchez-Flores, A., 2015. The road to metagenomics: from microbiology to DNA sequencing technologies and bioinformatics. Front. Genet. 6, 348. Available from: http://www.ncbi.nlm.nih.gov/pubmed/26734060.

Handelsman, J., 2004. Metagenomics: application of genomics to uncultured microorganisms. Microbiol. Mol. Biol. Rev. 68 (4), 669–685. Available from: http://www.ncbi.nlm.nih.gov/pubmed/15590779.

Hesse, W., Gröschel, D.H.M., 1992. Walther and angelina hesse-early contributors to bacteriology. ASM 58 (8), 425–428.

Hill, T.C.J., et al., 2003. Using ecological diversity measures with bacterial communities. FEMS Microbiol. Ecol. 43 (1), 1–11.

Huttenhower, C., et al., 2012. Structure, function and diversity of the healthy human microbiome. Nature 486 (7402), 207–214. Available from: http://www.ncbi.nlm.nih.gov/pubmed/22699609.

Janda, J.M., Abbott, S.L., 2007. 16S rRNA gene sequencing for bacterial identification in the diagnostic laboratory: pluses, perils, and pitfalls. J. Clin. Microbiol. 45 (9), 2761–2764. Available from: http://www.ncbi.nlm.nih.gov/pubmed/17626177.

Li, J., et al., 2014. An integrated catalog of reference genes in the human gut microbiome. Nat. Biotechnol. 32 (8), 834–841. Available from: http://www.nature.com/doifinder/10.1038/nbt.2942.

Lloyd-Price, J., Abu-Ali, G., Huttenhower, C., 2016. The healthy human microbiome. Genome Med. 8 (51).

Marchesi, J.R., Ravel, J., 2015. The vocabulary of microbiome research: a proposal. Microbiome 3 (1), 31. Available from: http://microbiomejournal.biomedcentral.com/articles/10.1186/s40168-015-0094-5.

Marchesi, J.R., et al., 2016. The gut microbiota and host health: a new clinical frontier. Gut 65 (2), 330–339. Available from: http://www.ncbi.nlm.nih.gov/pubmed/26338727.

Methé, B.A., et al., 2012. A framework for human microbiome research. Nature 486 (7402), 215–221. Available from: http://www.nature.com/doifinder/10.1038/nature11209.

Moeller, A.H., et al., 2016. Cospeciation of gut microbiota with hominids. Science 353 (6297), 380–382.
National Institutes of Health, 2012. NIH Human Microbiome Project defines normal bacterial makeup of the body. Available from: https://www.nih.gov/news-events/news-releases/nih-human-microbiome-project-defines-normal-bacterial-makeup-body.
NIH HMP Working Group, et al., 2009. The NIH Human Microbiome Project. Genome Res. 19 (12), 2317–2323. Available from: http://www.ncbi.nlm.nih.gov/pubmed/19819907.
Pace, N.R., Sapp, J., Goldenfeld, N., 2012. Phylogeny and beyond: scientific, historical, and conceptual significance of the first tree of life. Proc. Natl. Acad. Sci. U. S. A. 109 (4), 1011–1018. Available from: http://www.ncbi.nlm.nih.gov/pubmed/22308526.
Qin, J., et al., 2010. A human gut microbial gene catalog established by metagenomic sequencing. Nature 464 (7285), 59–65. Available from: http://www.nature.com/doifinder/10.1038/nature08821.
Qin, J., et al., 2012. A metagenome-wide association study of gut microbiota in type 2 diabetes. Nature 490 (7418), 55–60. Available from: http://www.nature.com/doifinder/10.1038/nature11450.
Robles-Alonso, V., Guarner, F., 2014. From basic to applied research. J. Clin. Gastroenterol. 48, S3–S4. Available from: http://www.ncbi.nlm.nih.gov/pubmed/25291122.
Schnupf, P., et al., 2015. Growth and host interaction of mouse segmented filamentous bacteria in vitro. Nature 520 (7545), 99–103. Available from: http://www.nature.com/doifinder/10.1038/nature14027.
Sender, R., Fuchs, S., Milo, R., 2016. Revised estimates for the number of human and bacteria cells in the body. PLoS Biol. 14 (8), e1002533. Available from: http://www.ncbi.nlm.nih.gov/pubmed/27541692.
Shreiner, A.B., Kao, J.Y., Young, V.B., 2015. The gut microbiome in health and in disease. Curr. Opin. Gastroenterol. 31 (1), 69–75.
Staley, J.T., Konopka, A., 1985. Measurement of in situ activities of nonphotosynthetic microorganisms in aquatic and terrestrial habitats. Annu. Rev. Microbiol. 39 (1), 321–346. Available from: http://www.annualreviews.org/doi/10.1146/annurev.mi.39.100185.001541.
Turnbaugh, P.J., et al., 2007. The human microbiome project. Nature 449 (7164), 804–810. Available from: http://www.nature.com/doifinder/10.1038/nature06244.
Ursell, L.K., et al., 2012. Defining the human microbiome. Nutr. Rev. 70 (Suppl. 1), S38–S44. Available from: http://www.ncbi.nlm.nih.gov/pubmed/22861806.
Whitman, W.B., Coleman, D.C., Wiebe, W.J., 1998. Prokaryotes: the unseen majority. Proc. Natl. Acad. Sci. U. S. A. 95 (12), 6578–6583. Available from: http://www.ncbi.nlm.nih.gov/pubmed/9618454.
Woese, C.R., 1987. Bacterial evolution. Microbiol. Rev. 51 (2), 221–271. Available from: http://www.ncbi.nlm.nih.gov/pubmed/2439888.
Woese, C.R., Fox, G.E., 1977. Phylogenetic structure of the prokaryotic domain: the primary kingdoms. Proc. Natl. Acad. Sci. U. S. A. 74 (11), 5088–5090. Available from: http://www.ncbi.nlm.nih.gov/pubmed/270744.
Woese, C.R., Kandler, O., Wheelis, M.L., 1990. Towards a natural system of organisms: proposal for the domains Archaea, Bacteria, and Eucarya. Proc. Natl. Acad. Sci. U. S. A. 87 (12), 4576–4579. Available from: http://www.ncbi.nlm.nih.gov/pubmed/2112744.
Zilber-Rosenberg, I., Rosenberg, E., 2008. Role of microorganisms in the evolution of animals and plants: the hologenome theory of evolution. FEMS Microbiol. Rev. 32 (5), 723–735. Available from: http://www.ncbi.nlm.nih.gov/pubmed/18549407.

CHAPTER 2

The Gut Microbiota

Objectives

- To understand the specific importance of the microorganisms residing in the gut.
- To become familiar with the major factors that influence bacterial colonization throughout the digestive tract and with what is known about bacterial composition at each site.
- To gain perspective on the complex role of the gut immune system and how it is set up to defend against pathogenic bacteria.
- To learn about the gut-brain axis and the potential influence of gut microbiota on neuronal activation, endocrine signals, and immune signals that impact the brain.
- To become acquainted with other gut microorganisms that might be relevant in health and disease.

BACTERIA IN THE DIGESTIVE TRACT

The **digestive tract**, in simple terms, is a large tube of muscle through which ingested food moves. It encompasses the mouth, pharynx, esophagus, stomach, small intestine, and large intestine, ending with the rectum and anus (see Fig. 2.1).

In addition to the enormous population of bacteria in and on the human body (an estimated 39 trillion bacterial cells), there exist a substantial number of viruses of both eukaryotes and prokaryotes, archaea, and fungi (Reinoso Webb et al., 2016). Of all the sites in and on the body, the digestive tract or **gut** houses the most dense, diverse, and dynamic collection of microorganisms (Methé et al., 2012). Within the gut environment (and in particular the colon), microorganisms interact with body systems, including the nervous, immune, and endocrine systems.

Gut microorganisms are associated with both the space inside the digestive tract, known as the **lumen**, and the innermost layer of the digestive tract, called the **mucosa**. The microbial content of the gastrointestinal (GI) tract changes along its length, ranging from a narrow diversity and low numbers of microbes in the esophagus and stomach to a wide diversity and high numbers in the colon. Although "gut microbiota" properly refers

Gut Microbiota
https://doi.org/10.1016/B978-0-12-810541-2.00002-6

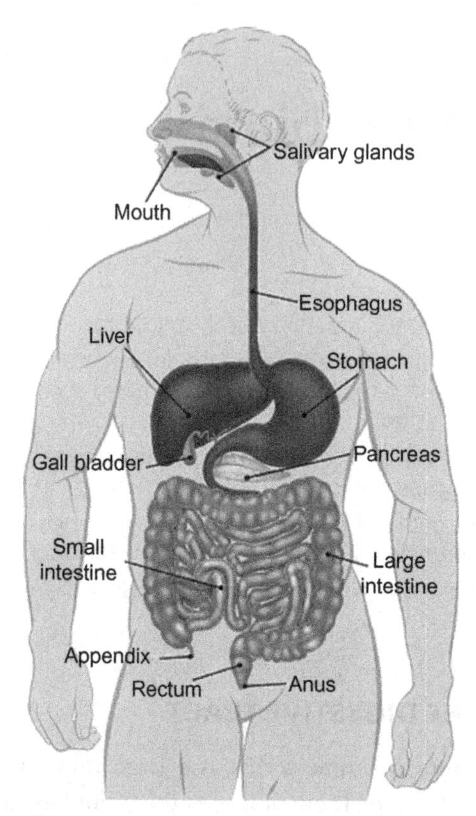

Fig. 2.1 Overview of the human digestive tract, from top (oral cavity) to bottom (rectum/anus). *(Reproduced with permission from Thinkstock. Human Digestive System Tract by ChrisGorgio.)*

to the microorganisms throughout the digestive tract, the term usually denotes the colonic fecal microbiota, since the microorganisms at this site are the most well studied and appear to have particular relevance to health. Over 90% of bacteria in the adult human intestinal tract are members of the phyla Bacteroidetes and Firmicutes, but members of Proteobacteria, Actinobacteria, Fusobacteria, Verrucomicrobia, and Cyanobacteria are also observed (Lozupone et al., 2012).

Factors Influencing Gut Bacterial Colonization

The ebb and flow of available nutrients affects the microbes that populate the gut (as discussed further in Chapter 6), but factors determining the composition of the microbiota throughout the GI tract are complex: the bacterial communities differ at each site in part because of local variables that influence colonization. Oxygen content is one such variable; the anatomy and physiology of the intestine create the conditions for steep differences in oxygen concentrations, with near-anoxic conditions at the luminal midpoint (Espey, 2013). The upper part of the GI tract, being open to the external environment, is aerobic, and thus, microorganisms capable of growing in the presence of oxygen tend to colonize this region. Of the three classes of microbes when it comes to oxygen relationships—**obligate aerobes** (requiring oxygen), **facultative anaerobes** (growing optimally in oxygen but not requiring it), and **aerotolerant anaerobes** (not metabolizing oxygen but able to survive in its presence)—the latter two form the bulk of GI tract microbes. Common bacteria like *Streptococcus* and *Lactobacillus* species (spp) are examples of aerotolerant anaerobes.

The pH—acidity or alkalinity—of a GI tract site is another factor that influences which bacteria thrive there. Each bacterial species prefers a certain pH range, and the community as a whole is shaped by this variable. As shown in bacterial ecosystems outside the human body, pH can be highly predictive of bacterial community diversity at a given site (Fierer and Jackson, 2006).

The entire GI tract is ringed with circular muscles that create waves of muscular contraction called **peristalsis**, which continuously push the intestinal contents further along. Peristalsis occurs at varying rates and intensities, depending on the presence of food and the location in the GI tract. Evidence from animal models suggests motility in the GI tract of the host may influence gut microbial community dynamics (Wiles et al., 2016), although more studies are required for elucidating mechanisms in humans.

Gut Microbiota and Digestion

Digestion is the process involving transport of food through the GI tract, breakdown of food into absorbable components, absorption, and elimination of solid waste. A growing body of scientific evidence is now revealing digestion as a complex series of mechanical and chemical processes that are under the influence of neurological and hormonal signals. Microbial activities in the human GI tract are emerging as key players in these digestive activities.

The journey of food from the top end of the digestive tract to the bottom end, and what is known about the bacterial community at each site are described in detail below.

Oral Cavity

The journey of food through the digestive tract begins in the oral cavity (mouth), which contains the teeth and tongue. Chewing, or mastication, breaks down large pieces of food into smaller ones, whereas the tongue moves food around to facilitate chewing and initiates the movements that enable it to be swallowed. Salivary glands secrete saliva that lubricates the food and supplies enzymes that digest carbohydrates, lipids, and proteins; alpha-amylase, for example, begins digesting the carbohydrates starch and glycogen upon contact with saliva (Janson and Tischler, 2012). Saliva also traps molecules produced by normal oral bacteria that add to the taste sensation of food compounds (Janson and Tischler, 2012). Food is chewed and formed into a soft, ball-like mixture called a **bolus**, and is ready for swallowing.

The oral microbiome is relatively well studied. In healthy adults, several distinct habitats with unique sets of bacteria exist in the mouth: for example, different communities exist on the teeth, gingival sulcus (the space between the tooth and the surrounding gum tissue), tongue, cheeks, hard and soft palates, and tonsils. Around 700 species have been detected in the human oral microbiota, although each individual may harbor only a few hundred of these species (Kilian et al., 2016).

Data from the Human Microbiome Project showed the oral microbiota of healthy individuals included both pathogens and bacteria from the outside environment. Bacteria from the families Porphyromonadaceae, Veillonellaceae, and Lachnospiraceae were common to all oral regions, although the genera in these families were distributed differently at each particular site in the mouth (Segata et al., 2012).

Salivary proteins and gingival crevicular fluid (GCF) act as nutrients for the growth of certain microorganisms (van 't Hof et al., 2014), while glycoproteins adhere to tooth enamel and form the substrate on which oral bacteria like *Streptococcus* spp. form a biofilm called dental plaque. Salivary compounds like hydrogen peroxide (H_2O_2), a highly reactive oxygen-containing molecule, can harm bacteria. Furthermore, saliva contains several **antimicrobial substances**—lactoferrin, salivary lactoperoxidase, lysozyme, and thiocyanate—which can damage bacteria (Janson and Tischler, 2012). Nitrite, converted by oral bacteria from dietary nitrates, is another compound in saliva with antimicrobial effects; when reduced, it produces nitric oxide, which inhibits the growth of some bacteria (Doel et al., 2004).

An increase in microbiota complexity accompanies the initiation of both periodontal disease (affecting the oral soft tissues and bone supporting the teeth) and caries (affecting the dental hard tissues) (Costalonga and Herzberg, 2014). In periodontitis, several species of protein-degrading bacteria, including the keystone pathogen *Porphyromonas gingivalis*, are increased, while the initiation of dental caries has been linked to the sugar-fermenting bacterium *S. mutans* and possible others. Yet, oral microbiome data suggest that a single-pathogen model cannot account for either caries or periodontitis: in these diseases, many constituents of the bacterial community are perturbed (Costalonga and Herzberg, 2014).

Pharynx

When swallowed, food passes through the pharynx (throat), a short tube shared by both the digestive and respiratory systems. A flap of cartilage in the throat, called the epiglottis, temporarily closes off the trachea (airway) to prevent choking during the act of swallowing.

Little is known about the human pharyngeal microbiota, but a small study of six healthy individuals found changes in the throat microbiota after infection with human rhinovirus, which causes the common cold. The five most abundant genera preinfection were *Streptococcus*, *Prevotella*, *Rothia*, *Veillonella*, and *Haemophilus*; after infection, the researchers observed increases in relative abundances of certain species: *Haemophilus parainfluenzae* and *Neisseria subflava*, as well as *Staphylococcus aureus* (Hofstra et al., 2015).

Esophagus

The esophagus is the tube leading down to the stomach. At each end of the esophagus is a sphincter—a circular muscle that is able to open and close. After food moves through the esophagus into the stomach, the lower esophageal sphincter closes to prevent its movement back into the esophagus.

The challenge of studying the esophageal microbiota is due to the need for invasive sampling procedures and to the dynamic nature of the esophageal environment. At any moment, pH values and microbial communities are affected by both saliva and the reflux of gastric material into the esophagus. Members of the genus *Streptococcus* appear to be dominant in the microbiota of the healthy esophagus (Di Pilato et al., 2016), although the presence of several other taxa, including *Prevotella* and *Veillonella*, has also been reported (Pei et al., 2004); the bacterial composition seems highly similar to that of the oral cavity (Snider et al., 2016).

Stomach

A muscular sac that grinds and churns food, the stomach adds acid and enzymes to the bolus, forming a semiliquid called **chyme**. Under the control of the autonomic nervous system (the part of the nervous system that controls the function of internal organs) and hormones, the stomach secretes hydrochloric acid, various digestive enzymes (notably pepsinogen), mucus, and other gastric juices (Janson and Tischler, 2012). The stomach functions as a short-term storage place for food before it continues through the digestive tract. Below the stomach, a circular muscle called the pyloric sphincter controls the flow of partially digested food out of the stomach and into the small intestine.

The known acidic environment of the stomach initially led people to assume that few microorganisms could live there. But evidence now demonstrates that the human stomach houses a microbial ecosystem distinct from other digestive tract sites, which is dependent on how gastric microbiota species either resist or make use of acid (Nardone and Compare, 2015). Studies of gastric biopsies show the microbial ecosystem of the stomach is dominated by the phyla Proteobacteria, Firmicutes, Actinobacteria, Bacteroidetes, and also Fusobacteria (Bik et al., 2006; Maldonado-Contreras et al., 2011). The gastric microbiota composition is dynamic and at any moment the stomach may house a variety of transient species originating in the oral cavity, possibly as a result of swallowing (Nardone and Compare, 2015).

One stable colonizer of the stomach, which is present in between one-third and two-thirds of the global population (Eusebi et al., 2014) with a median prevalence around 50%, is the bacterium *Helicobacter pylori*. Most individuals show changes in the structure of the gastrointestinal microbiota—with less diversity in general—after colonization with this bacterium. Mechanisms by which this bacterium could alter the microbiota include perturbing the gastric environment, inducing hormone secretion, and changing inflammatory response (He et al., 2016). For example, *H. pylori* infection leads to an increase of the gastric pH over the long term, perhaps allowing transient bacteria to increasingly colonize the stomach (Nardone and Compare, 2015). Gastric *H. pylori* colonization may also significantly influence the duodenal and oral communities (Schulz et al., 2016).

Small Intestine

The small intestine, ~5–6 m in length, is the most important site in the human digestive tract for digesting food and absorbing nutrients. Most of the human host's enzymatic digestion and absorption of nutrients (in particular, lipids and simple carbohydrates) takes place in the small intestine. This part of the digestive tract comprises three sections: the duodenum (top portion), the jejunum (middle portion), and the ileum (lower portion that attaches to the colon). The ileocecal valve is located at the juncture of the ileum and the colon.

The microbiota of the small intestine is difficult to sample, but in the few available studies, researchers have found the microbes in this region less populous than in other areas of the digestive tract, and highly dynamic, seemingly amenable to modulation by diet (El Aidy and van den Bogert, 2015). *Streptococcus* and *Veillonella* are important genera of commensals in the small intestine.

Large Intestine

Below the small intestine lies the large intestine. The 1.5 m-long tube called the colon is the location for continued water reabsorption, uptake of microbe-derived vitamins, and stool formation (Janson and Tischler, 2012). The rectum is the muscular end part of the colon, which extends down to the opening that is the terminus of the digestive tract, the anus. Through these structures, stool is excreted.

The colon is the densest site of microbes in the entire body and by far the most studied. Between 300 and 1000 different species of bacteria live in the large intestine of healthy adults, primarily from the phyla Bacteroidetes, Firmicutes, and Proteobacteria (National Institutes of Health, 2012; Qin et al., 2010).

Researchers often use the composition of fecal microbiota as a shorthand for that of the luminal and/or mucosal environments in the colon. Yasuda and colleagues undertook a study of the biogeography of the mucosal, luminal, and stool microbiota in monkeys and found the stool microbiota is indeed a good proxy for that of the colonic lumen and mucosa; surprisingly, it also correlated well with the small intestinal microbiota (except when it came to small intestinal Proteobacteria, which were underdetected in the stool). The group found a slight enrichment of facultative anaerobes in the mucosa (*Helicobacter*, for example, was highly enriched) and of obligate anaerobes (e.g., short-chain fatty acid producers) in the lumen (Yasuda et al., 2015).

An emerging hypothesis about the **appendix**—a narrow sac, hanging off the colon, which serves as a storage place for lymph cells—identifies it as a protective niche for beneficial gut bacterial species. The appendix of a healthy individual harbors a microbiota distinct from that of the feces; increased *Fusobacteria* spp., for example, appear to be present (Rogers et al., 2016). A recent analysis of 533 mammalian species suggested the appendix may have a function related to adaptive immunity, since species that had an appendix tended to show higher concentrations of immune tissue (i.e., **lymphoid tissue**) in the **cecum**, which is a pouch at the junction of the small and large intestines (Smith et al., 2017).

Importantly, the colon is the site of breakdown for materials that escape digestion earlier on in the digestive tract—namely, dietary fiber, resistant starches, and noncarbohydrate substrates. Colonic bacteria produce enzymes that break up (ferment) these materials; among the metabolic waste products from this process is an important group of metabolites called **short-chain fatty acids** (SCFAs), which include **acetate**, **propionate**, and **butyrate**. Some SCFAs are excreted, but most (95%) are used by colonic cells as a source of energy. SCFAs provide about 10% of total daily calories for the host (Duncan et al., 2007), but their role in metabolism is incompletely understood at present: in particular, the apparent paradox that they are associated with metabolic health benefits and appear to protect against obesity while also providing calories that could contribute to obesity (Boulangé et al., 2016). SCFAs serve as important messengers between the microbiota and the immune system, playing a role in the development and function of intestinal epithelial cells and leukocytes (Corrêa-Oliveira et al., 2016). The health effects of SCFAs may depend on the delicate balance between production, uptake, and excretion; this issue is discussed further in Chapter 4.

Bacteria in the colon serve the additional function of increasing the body's absorption of remaining lipids, proteins, and minerals such as calcium, magnesium, and iron (Janson and Tischler, 2012). Colonic microbiota can produce vitamins such as thiamine, riboflavin, niacin, biotin, pantothenic acid, and folate (i.e., B vitamins), as well as vitamin K (Biesalski, 2016); while vitamins supplied in the diet are absorbed in the small intestine, microbe-produced vitamins are absorbed in the colon. The colonic bacteria also produce secondary bile salts, which may be passively absorbed in the colon or excreted in the feces.

Many disease states are associated with disturbances in the colonic mucosal or luminal/fecal microbiota. These are covered in more detail in Chapter 4.

Other Organs: Liver, Gallbladder, Pancreas, and Spleen

Besides the ones described above, several other organs are also important for normal digestion: the liver, gallbladder, pancreas, and spleen. These organs do not appear to house a discernibly important microbiota.

The liver is a complex organ with a huge number of functions, including those implicated in metabolism and detoxification. When it comes to digestion, the liver makes and secretes bile for fat digestion and absorption; stores vitamins, sugars, fats, and other nutrients; regulates hormones; and metabolizes foreign compounds that are synthesized or modified by gut microbes. The liver also prevents live microbes from entering the blood and contributes to immunologic activity based on signals from microbial metabolites. Based on these functions, Macpherson, Heikenwalder, and Ganal-Vonarburg recently proposed the liver as a "nexus" for establishing and maintaining mutual dependence between host and microbes (Macpherson et al., 2016).

The gall bladder is a digestive organ that stores and concentrates bile. When the gall bladder receives the signal that fat is present in the duodenum (i.e., from dietary intake), it contracts to release the bile needed for fat digestion and absorption.

The pancreas is a gland that secretes digestive enzymes (carbohydrases, lipases, nucleases, and proteolytic enzymes) and secretions into the duodenum when it receives the hormonal signal that food is present. It also releases hormones into the blood to help with glucose homeostasis.

Finally, the spleen is an immune system organ that primarily acts as a blood filter; it plays a minor role in digestion because of its connection to the blood vessels of the stomach and pancreas.

Gut Microbiota and the Immune System

Because they are constantly exposed to large populations of microorganisms, the mucosal tissues of the respiratory, gastrointestinal, and genitourinary tracts represent the most likely portals of entry for microbial pathogens. These tissues are consequently protected by sophisticated physical and immunologic barriers. The intestinal epithelium of the small and large intestines (informally, the **gut barrier**) is especially vulnerable to infection because it is represented by only a single layer of cells—this expedites uptake of nutrients, water, and electrolytes. Moreover, to further facilitate these processes, the luminal surface of the epithelium contains numerous projections or **villi** to increase the epithelial surface area to an estimated 32 m^2, with all but 2 m^2 representing the small intestine (Helander and Fandriks, 2014). In comparison, the surface area of the skin on an average human is 1.5–2 m^2 or about the same as the colon. Thus, protecting the large area of the small intestine from possibly harmful microorganisms is especially challenging, and not surprisingly, many of the body's immune defenses are concentrated in this region. The GI tract's immune system in this area is known as the **gut-associated lymphoid tissue (GALT)**.

The intestinal epithelium is protected from bacteria by several lines of defense, as discussed further below: (1) an outer mucus layer, (2) epithelial cell secretion of different antimicrobial peptides, (3) Paneth cell activity, and (4) synthesis and secretion of immunoglobulin A (Reinoso Webb et al., 2016).

GALT Organization and Function

Cells adjacent to each other in the intestinal epithelium are secured to each other by transmembrane multiprotein complexes, forming a selectively permeable seal between the cells (Lee, 2015). This arrangement, called the **tight junction**, represents a significant component of gut barrier function. The intestinal epithelium is composed of several different types of cells. **Enterocytes**, the most numerous cell type, are primarily responsible for nutrient transport. These host cells produce membrane-bound **mucins** (large, heavily glycosylated proteins) that extend into the intestinal lumen, forming a structure known as a **glycocalyx** (an outer viscous cover), which may serve as a localized protective covering for the luminal enterocyte membrane (Johansson and Hansson, 2016). The second in abundance are the **epithelial goblet cells**, which are the major producers of intestinal **mucus**. They secrete mucins, which are not bound to the enterocyte membrane, into the luminal side of the epithelium. The large water-holding capacity of the mucins' oligosaccharide components gives them their gel-like consistency, and their adherence to the epithelium creates a significant physical antimicrobial barrier (although a complement of microbes also reside in this outer layer of loosely adherent mucus; Li et al., 2015). A third principal epithelial cell type is the **Paneth cells**, which occupy a strategic location in the crypts of the small intestinal epithelium and play a significant role in gut innate immunity (Clevers and Bevins, 2013). They are cytologically recognizable by their large number of intracellular secretory granules that contain several antimicrobial peptides, such as the alpha-defensins, that play key roles in controlling mucosal tissue colonization and protection against host infection. For example, alpha-defensins bind to bacterial cell membranes and make them permeable, killing the microorganism (Bevins, 2013). Every 2–5 days, the gut epithelial cells are replaced, and new replacement cells are derived from stem cells found in the intestinal crypts. The crypt-localized Paneth cells may also be important in protecting these precious stem cells. Finally, although they constitute only about 1% of the total epithelial cells, **enteroendocrine cells** are present, which form the largest endocrine system in the body (Moran et al., 2008). In response to specific stimuli, they secrete a variety of hormones that control an assortment of digestive tract functions through the enteric nervous system, ranging from the control of food uptake to mucosal immunity.

A layer of connective tissue called the **lamina propria** is located below the gut epithelium. The lamina propria is home to several types of organized lymphoid tissues involved in gut adaptive immunity and to a large population of immune cells ranging from macrophages to various classes of lymphocytes. **Antibodies** are immune proteins produced to counteract specific "foreign" substances in the body, and in humans, about 80% of the total antibody-producing plasma cells are found in the gut immune system (Gommerman et al., 2014). There are several pathways (i.e., series of consecutive reactions) for the generation of mucosal antibodies, but the major site for induction of the gut adaptive immune response is the lymphoid tissue called the **Peyer's patch**. About 100–200 Peyer's patches occur along the small intestine in the regions of the jejunum and especially the ileum (Reboldi and Cyster, 2016). The organization of a Peyer's patch is similar to a lymph node (see Fig. 2.2A). Within the subepithelial dome of the Peyer's patch, there are follicles (B-cell regions) and interfollicular regions occupied by T cells. The gut epithelium that overlies the Peyer's patch is called the **follicle-associated epithelium** (FAE). About 10% of the cells in the FAE are unique **microfold (M) cells** (Mabbott et al., 2013) (see Fig. 2.2B). Unlike other gut epithelial cells, the surfaces of M cells are relatively free of mucin and directly accessible to microorganisms and other materials in the gut lumen. M cells are notable for their ability to phagocytose these antigenic (foreign) materials from the lumen and deliver them by transcytosis to a unique pocket on their basolateral membrane (Fig. 2.2B). There, the antigens are delivered to antigen-presenting cells such as lymphocytes, macrophages, and dendritic

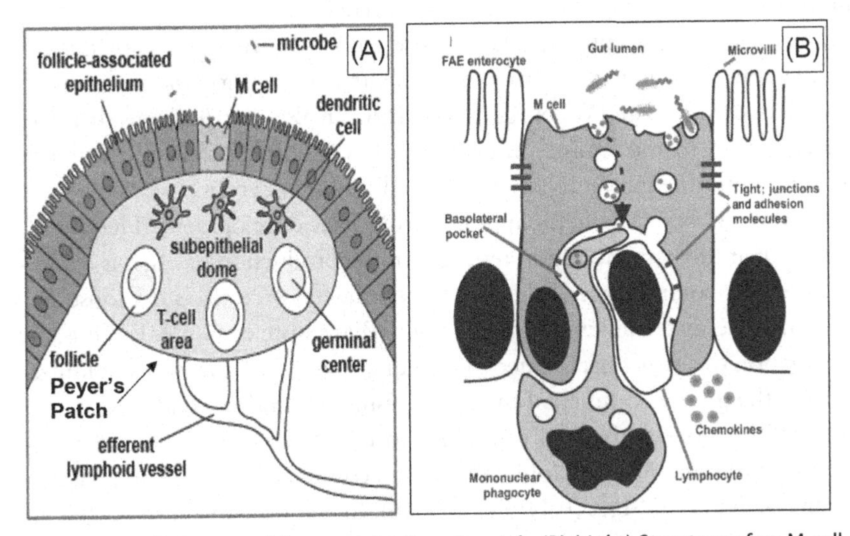

Fig. 2.2 (A) (left) Structural features of a Peyer's patch. (B) (right) Structure of an M cell, flanked by enterocytes that are only partially shown. Antigens from the gut lumen are transcytosed to a mononuclear phagocyte and a lymphocyte in the basolateral pocket of the M cell. The dark structures are cell nuclei. *(Reproduced with permission from (A) Gary E. Kaiser; (B) Mabbott, N. A. et al., 2013. Microfold (M) cells: important immunosurveillance posts in the intestinal epithelium. Mucosal Immunol. 6(4), 666–677.)*

cells that have been attracted to the M-cell pocket by chemokines (signaling proteins) that are constitutively secreted by the FAE. Antigen-activated B cells, under instructions from antigen-activated helper T cells, enter germinal centers of Peyer's patches where they undergo stages of differentiation, three of which are noteworthy (see Fig. 2.2A). First, class switch recombination is a process that results in commitment to produce antibodies of the class known as **immunoglobulin A (IgA)**. This event is unique to mucosal immunity, since, for example, the predominant antibody class in the systemic immune system is IgG. Second, somatic hypermutation is a process that results in increasing the affinity of IgA for the antigen that activated B-cell differentiation. Third, germinal center B cells acquire gut-specific homing mechanisms such as gut-specific adhesion molecules and chemokine receptors (Mora et al., 2006). The activated B cells leave the Peyer's patch, enter the circulation, and eventually return to the lamina propria utilizing their acquired gut-homing mechanisms. Upon returning to the lamina propria, the B cells complete their differentiation and become functional IgA-secreting plasma cells. The IgA is secreted almost exclusively as a dimer—a two-part chemical compound (Gommerman et al., 2014). Dimeric IgA, known as **secretory IgA** or **SIgA**, is transported across the gut epithelium into the gut lumen by transcytosis mediated by a protein called polymer immunoglobulin receptor (pIgR) on the basolateral membrane of the gut epithelium. SIgA protects the mucosa by neutralizing microorganisms or their products (Strugnell and Wijburg, 2010). SIgA can also bind antigens that have gained access to the lamina propria, and these complexes are subsequently exported into the gut lumen by pIgR-mediated transcytosis.

Although Peyer's patches appear during fetal development (Heel et al., 1997), their further maturation resulting in the production of SIgA is dependent on the colonization of the gut by commensal bacteria as demonstrated in human neonates and in germ-free mice (Benveniste et al., 1971). Thus, the maturation of the neonatal mucosal immune system and the initial colonization of the gut by commensal microorganisms go hand in hand; a reciprocal interaction between the two is thought to be crucial in ultimately defining the composition of the microbiota. This process depends on the mucosal protective activities of SIgA (Pabst et al., 2016). In addition, the enterocytes and immune cells in the lamina propria express a variety of receptors that can detect microbial products and respond in an appropriate way (Thaiss et al., 2016). It is important to note that these mechanisms are designed to maintain homeostasis between the resident microbiota and the gut immune system and therefore do not exclusively target pathogens. Further details on how gut microbiota commensals and symbionts regulate the immune system and how they fail to do so in certain disease states appear in Chapter 4.

Gut Microbiota and Brain Structure/Function

In the study of gastroenterology, a specialty called "neurogastroenterology" is concerned with neural influences on digestive function. The concept of a **gut-brain axis**, which, in a 2006 article (Jones et al., 2006), was defined as "the combined functioning of GI intestinal motor, sensory and central nervous system (CNS) activity," is not new (Track, 1980), but only recently are scientists discovering that gut microbiota plays a key role in the two-way communication between the gut and the brain.

The CNS and parts of the peripheral nervous system (PNS) in humans carry signals between the digestive tract and the brain. The brain and spinal cord comprise the CNS; scientists have long known that the brain's ~100 billion neurons, interconnected in complex circuits, collect information from all parts of the body and produce a large range of motor outputs enabling humans to survive and successfully interact with the environment—from breathing and walking to chewing and driving. The role of the CNS in controlling digestion, under normal circumstances, is a minor one. The CNS helps control stomach contractions and acid secretion through vagal reflex circuits and helps regulate intestinal permeability and secretion of mucus; however, cutting the vagus nerve in animals and humans has only a minor effect on gastrointestinal function (Fossmark et al., 2013).

A subdivision of the PNS—the autonomic nervous system (ANS)—controls essential but involuntary body functions such as breathing and heartbeat, and is itself split into parasympathetic and sympathetic divisions. Forming part of the ANS but functioning as its own separate system is the **enteric nervous system** (ENS). The 200–600 million neurons of the ENS are spread out, weblike, along the entire length of the gut. ENS sensory neurons that have their cell bodies in the intestinal wall, called intrinsic primary afferent neurons, are in a prime position to respond to chemical and mechanical stimuli from the gastrointestinal tract. Meanwhile, ENS motor neurons act on cells including epithelial tissue, mucosal glands, smooth muscle, and blood vessels, and also affect the immune and endocrine cells distributed along the digestive tract (Costa et al., 2000). Thus, the ENS plays a role in digestion not only by controlling motor functions, local blood flow, and mucosal transport and secretions, but also by helping modulate immune and endocrine functions (Costa et al., 2000).

The gut microbiota appears to play a role in shaping brain function and behavior (and perhaps even brain structure), but mechanisms are difficult to uncover in humans. The majority of the research to date involves animal models, as described below. See Fig. 2.3 for an overview of the ways in which gut microbes may influence the brain in humans.

The Brain in the Absence of a Microbiota

Germ-free rodents provide preliminary evidence of the influence of gut microbiota on brain structure and function via the gut-brain axis. First, germ-free mice show abnormalities in brain function (Luczynski et al., 2016): an abnormal response to stressful situations, different patterns of exploratory behaviors and social interaction, and changes in cognition. These animals that lack a microbiota also have physically observable differences in their

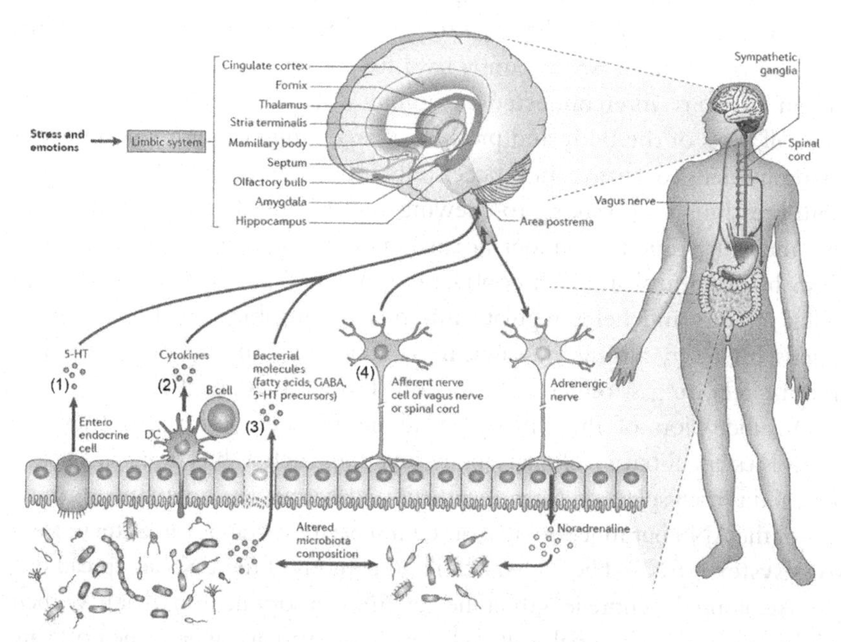

Fig. 2.3 The potential influence of intestinal microbiota in the human gut-brain axis. The gut microbiota may influence the brain through several pathways: (1) the release of gut hormones like serotonin (5-HT) from enteroendocrine cells, (2) cytokine release from mucosal immune cells (e.g., dendritic cells or DC), (3) bacterial products (e.g., gamma-aminobutyric acid or GABA) gaining access to the brain via the bloodstream and area postrema, and (4) afferent neural pathways (such as the vagus nerve). Stress may influence gut microbial composition either by altering stress hormones or sympathetic neurotransmitters or by affecting hormones that change bacterial signaling and thus modulate microbiota composition. *(Reproduced with permission from Collins, S. M., Surette, M., Bercik, P., 2012. The interplay between the intestinal microbiota and the brain. Nat. Rev. Microbiol. 10(11), 735–742. Macmillan Publishers Ltd. Copyright (2012).)*

brains. One physical difference is that their microglia (cells that constitute active immune defense in the brain) are defective and immature (Erny et al., 2015); furthermore, they show hypermyelinated axons in the prefrontal cortex, a difference associated with a shorter lifespan (Hoban et al., 2016). Germ-free mice also have defects in the development of synapses between neurons (Diaz Heijtz et al., 2011): they show alterations in the production of new synapses and the pruning of existing ones. And germ-free mice or those with a severely disrupted microbiota also show differing expression of the neuromodulator BDNF (brain-derived neurotrophic factor), a protein that influences cognition by promoting maturation and survival of developing neurons and maintenance of mature neurons (Bercik et al., 2011). However, it is unknown at present how these results relate to humans.

Infuence of Microbiota on Neuronal Activation

During digestion, gut-lining cells gather information about activities in the digestive tract and convey messages to other cells in the gut wall (primarily endocrine cells); the messages travel to nearby sensory neurons, especially the vagus nerve, and continue upward to the brain (Perez-Burgos et al., 2014). The **vagus nerve** (cranial nerve X), which is activated by different kinds of signals as described below, is an important channel for transmitting information from the gut to the brain. Vagal signals travel bidirectionally, but 90% of signals travel in an upward direction, from gut to brain. Through this route, the brain receives from the gut a constant stream of information about digestive activities.

Animal studies show intestinal bacteria have the ability to change the excitability of neurons—that is, the threshold for evoking action potentials. For example, one study showed ENS myenteric neurons had reduced excitability in germ-free mice compared with normal mice. The excitability returned to normal when the mice were colonized with a microbiota (McVey Neufeld et al., 2013). Another study showed ingestion of the probiotic *L. reuteri* increased excitability of colonic neurons in rats (Kunze et al., 2009), while yet another found *L. rhamnosus* increased the firing rate of nerves in one region of the mouse brain—the mesenteric nerve bundle—but only when the vagus nerve was intact (Perez-Burgos et al., 2013).

Other probiotics appear to decrease excitability of ENS neurons. For example, a study of *Bifidobacterium longum* NCC3001 showed it could reduce the excitability of certain enteric sensory neurons in rats (Khoshdel et al., 2013). Together, these studies show different kinds of probiotic bacteria in the gut influence ENS neurons differently, either dampening or exciting them.

An additional way for microbes in the gut to influence ENS activity is through the production of molecules, like gamma-aminobutyric acid (GABA), which can act as local neurotransmitters. GABA made in the gut is separate from that made in the brain, but one study showed that in mice with a normal microbiota, *L. rhamnosus* (JB-1) in the gut could regulate receptors for GABA in several cortical regions and decrease the production of corticosterone in response to stress (Bravo et al., 2011)—and notably, this phenomenon only occurred if the vagus nerve was intact.

Influence of Microbiota on Endocrine Signals

Another, less direct, way that messages are relayed from the gut to the brain is via gut hormones and regulatory peptides (Zhou and Foster, 2015). **Serotonin** is a neurotransmitter biosynthesized in both the digestive tract and in the brain, with around 90% being produced in the gut. In 2015, Yano and colleagues found that microbe and host cells cooperate to make it: not only do specific bacteria in both humans and mice alter the metabolite signals that promote the production of gut serotonin, but also these bacteria appear to help regulate levels of serotonin in the colon and blood (Yano et al., 2015). The gut serotonin signaling system sends messages to the brain through the vagus nerve; indeed, vagal sensory signals can be activated by a range of hormones and other molecules. Cholecystokinin (CCK), glucagon-like peptide-1 (GLP-1), peptide YY (PYY), and ghrelin can all activate these signals, and evidence suggests they are very responsive to the nutrients encountered in the gut (Dockray, 2013). Enteroendocrine cells (which produce and release hormones) in the epithelial lining may be influenced by gut microbes (Uribe et al., 1994). For example, galanin—a peptide released from enteroendocrine cells that can trigger the release of corticotropin-releasing factor (CRF), which leads to increased cortisol—was found to be influenced by gut microbiota in rats, thus changing gut-brain communication (Tortorella et al., 2007).

Influence of Microbiota on Immune Signals

An additional means of messaging from the gut to the brain is through immune cells in the gut. These immune cells can release inflammation-inducing cytokines, which traverse the gut lining and reach the brain by activating the vagus nerve or by entering systemic circulation. Some limited evidence exists from animal models in support of the gut microbiota's effects on the immune responses that shape brain function: in one study, giving a probiotic mixture (VSL#3) to mice with liver inflammation resulted in reduced infiltration of certain white blood cells in the brain and reduced systemic immune activation, without any changes in disease severity or microbiota composition (D'Mello et al., 2015).

OTHER MICROORGANISMS IN THE DIGESTIVE TRACT

Bacterial members of the microbiota are almost certainly not the only relevant microorganisms in normal human body function, but very little is known about the other microbes' influence. Eukaryotes such as *Candida*, *Malassezia*, and *Saccharomyces* are pervasive in healthy populations (Underhill and Iliev, 2014), and interestingly, the microeukaryote *Blastocystis,* with a prevalence of around 20% in healthy populations (Andersen et al., 2015), is correlated with increased bacterial diversity and differences in the abundance of bacteria belonging to the genera *Ruminococcus* and *Prevotella* (Audebert et al., 2016). The virome, which is extensive in humans, could also be important since it is known that **bacteriophage** (viruses that infect bacteria), in their capacity as bacterial predators, can influence microbiota community structure. In terms of archaea, a small number of genera have been identified in the gut: species of the *Methanobrevibacter* genus are the most prevalent (Horz, 2015), with *M. smithii* in particular having distinct functions in the gut environment (Samuel et al., 2007). However, the molecular-profiling techniques for eukaryotes, viruses, and archaea are not as advanced as those for bacteria (Norman et al., 2014), so insights will depend on future advances in this realm.

Although the focus of the research to date (and thus this chapter) has been microbiota composition, growing evidence suggests microbiota function at each digestive tract site may be far more relevant to health and disease. It has been proposed that the gut microbiota must cover a core set of functions and that composition might vary as long as the genes encoding the required functions are present (Lloyd-Price et al., 2016). In the chapters that follow, important metabolomic research is described, which leads the way in identifying metabolites of gut bacteria that are active in signaling to multiple tissues and organs throughout the body.

Experimental Design and the Use of Animal Models

Although controlled manipulation of variables and empirical observation are the basis of scientific advancement, scientists studying human health often encounter situations where the precise variables that would yield a well-designed experiment are impractical or unethical. As a result, research that advances knowledge about human health must follow a more gradual course. Scientists must undertake, in sequence or in parallel, both experimental studies and observational studies (see Table 2.1). By using data from one type of research to make predictions for another type of research and adapting their conclusions to account for all results, scientists eventually gain insights into a particular phenomenon.

Table 2.1 Various scientific study designs

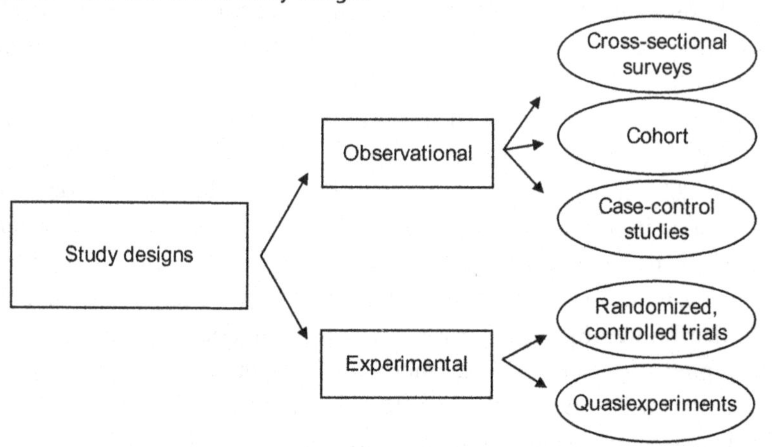

Data from different types of studies are taken together for insights about the gut microbiota in human health.

Consider the relationship between antibiotics and childhood obesity as an example of a phenomenon under study. The laboratory of Martin Blaser at New York University has found (1) experimental evidence from an animal model showing early-life antibiotics in specific dose patterns induce obesity in mice that is not corrected later in life (Cox et al., 2014) and (2) observational evidence from a human longitudinal birth cohort study that exposure to antibiotics before six months of age is associated with a higher body mass in childhood (Trasande et al., 2013). Continued convergence of evidence from the different types of studies could provide increased support for a causal relationship between antibiotic exposure and a higher body mass in childhood.

This example highlights another issue: further progress when it comes to the gut microbiota in human health depends greatly on suitable experimental models for elucidating mechanisms. Scientists have heavily relied on mouse models in gut microbiota research (Nguyen et al., 2015) for several reasons: mice, unlike humans, form homogeneous genetic populations because they are inbred—and this characteristic alone increases the probability of experimental reproducibility. The genetics of these mice are also better understood than the genetics of human populations. And experimental conditions are easier to control for mice, since humans cannot be maintained in cages. Furthermore, in mice, it is relatively easy to manipulate the gut microbiota in studies designed to examine the possible involvement of gut microbes in health and disease. Gnotobiotic mice have been popular models for over 50 years because all microbial exposures are known. But mouse models have significant limitations: mainly, they fail to represent all features of the comparable human disease. Mouse models pose particular problems when it comes to gut-brain axis research, since models of complex human-brain-related conditions like depression, Alzheimer's disease, and even autism are impossible to create.

REFERENCES

Andersen, L.O., et al., 2015. A retrospective metagenomics approach to studying *Blastocystis*. In: Marchesi, J. (Ed.), FEMS Microbiol. Ecol. 91 (7), 1–9. Available from: https://academic.oup.com/femsec/article-lookup/doi/10.1093/femsec/fiv072.

Audebert, C., et al., 2016. Colonization with the enteric protozoa Blastocystis is associated with increased diversity of human gut bacterial microbiota. Sci. Rep. 6, 25255. Available from: http://www.nature.com/articles/srep25255.

Benveniste, J., Lespinats, G., Salomon, J., 1971. Serum and secretory IgA in axenic and holoxenic mice article. J. Immunol. 100 (1), 1656–1662.

Bercik, P., et al., 2011. The intestinal microbiota affect central levels of brain-derived neurotropic factor and behavior in mice. Gastroenterology 141 (2), 599–609, 609–3. Available from: http://www.ncbi.nlm.nih.gov/pubmed/21683077.

Bevins, C.L., 2013. Innate immune functions of α-defensins in the small intestine. Dig. Dis. 31 (3–4), 299–304. Available from: http://www.ncbi.nlm.nih.gov/pubmed/24246978.

Biesalski, H.K., 2016. Nutrition meets the microbiome: micronutrients and the microbiota. Ann. N.Y. Acad. Sci. 1372 (1), 53–64.

Bik, E.M., et al., 2006. Molecular analysis of the bacterial microbiota in the human stomach. Proc. Natl. Acad. Sci. U. S. A. 103 (3), 732–737. Available from: http://www.ncbi.nlm.nih.gov/pubmed/16407106.

Boulangé, C.L., et al., 2016. Impact of the gut microbiota on inflammation, obesity, and metabolic disease. Genome Med. 8 (1), 42. Available from: http://genomemedicine.biomedcentral.com/articles/10.1186/s13073-016-0303-2.

Bravo, J.A., et al., 2011. Ingestion of Lactobacillus strain regulates emotional behavior and central GABA receptor expression in a mouse via the vagus nerve. Proc. Natl. Acad. Sci. U. S. A. 108 (38), 16050–16055. Available from: http://www.ncbi.nlm.nih.gov/pubmed/21876150.

Clevers, H.C., Bevins, C.L., 2013. Paneth cells: maestros of the small intestinal crypts. Annu. Rev. Physiol. 75, 289–311.

Corrêa-Oliveira, R., et al., 2016. Regulation of immune cell function by short-chain fatty acids. Clin. Transl. Immunol. 5 (4), e73. Available from: http://www.nature.com/doifinder/10.1038/cti.2016.17.

Costa, M., Brookes, S.J., Hennig, G.W., 2000. Anatomy and physiology of the enteric nervous system. Gut 47 (Suppl. 4), iv15–iv19, discussion iv26. Available from: http://www.ncbi.nlm.nih.gov/pubmed/11076898.

Costalonga, M., Herzberg, M.C., 2014. The oral microbiome and the immunobiology of periodontal disease and caries. Immunol. Lett. 162 (2 Pt A), 22–38. Available from: http://www.ncbi.nlm.nih.gov/pubmed/25447398.

Cox, L.M., et al., 2014. Altering the intestinal microbiota during a critical developmental window has lasting metabolic consequences. Cell 158 (4), 705–721. Available from: http://www.ncbi.nlm.nih.gov/pubmed/25126780.

Di Pilato, V., et al., 2016. The esophageal microbiota in health and disease. Ann. N.Y. Acad. Sci. Available from: http://doi.wiley.com/10.1111/nyas.13127.

Diaz Heijtz, R., et al., 2011. Normal gut microbiota modulates brain development and behavior. Proc. Natl. Acad. Sci. U. S. A. 108 (7), 3047–3052. Available from: http://www.ncbi.nlm.nih.gov/pubmed/21282636.

D'Mello, C., et al., 2015. Probiotics improve inflammation-associated sickness behavior by altering communication between the peripheral immune system and the brain. J. Neurosci. 35 (30), 10821–10830. Available from: http://www.ncbi.nlm.nih.gov/pubmed/26224864.

Dockray, G.J., 2013. Enteroendocrine cell signalling via the vagus nerve. Curr. Opin. Pharmacol. 13 (6), 954–958. Available from: http://www.ncbi.nlm.nih.gov/pubmed/24064396.

Doel, J.J., et al., 2004. Protective effect of salivary nitrate and microbial nitrate reductase activity against caries. Eur. J. Oral Sci. 112 (5), 424–428. Available from: http://doi.wiley.com/10.1111/j.1600-0722.2004.00153.x.

Duncan, S.H., et al., 2007. Reduced dietary intake of carbohydrates by obese subjects results in decreased concentrations of butyrate and butyrate-producing bacteria in feces. Appl. Environ. Microbiol. 73 (4), 1073–1078.

el Aidy, S., van den Bogert, B., 2015. The small intestine microbiota, nutritional modulation and relevance for health. Curr. Opin. Biotechnol. 32, 14–20.

Erny, D., et al., 2015. Host microbiota constantly control maturation and function of microglia in the CNS. Nat. Neurosci. 18 (7), 965–977. Available from: http://www.nature.com/doifinder/10.1038/nn.4030.

Espey, M.G., 2013. Role of oxygen gradients in shaping redox relationships between the human intestine and its microbiota. Free Radic. Biol. Med. 55, 130–140.

Eusebi, L.H., Zagari, R.M., Bazzoli, F., 2014. Epidemiology of Helicobacter pylori infection. Helicobacter (Suppl. 1), 1–5.

Fierer, N., Jackson, R.B., 2006. The diversity and biogeography of soil bacterial communities. Proc. Natl. Acad. Sci. U. S. A. 103 (3), 626–631. Available from: http://www.ncbi.nlm.nih.gov/pubmed/16407148.

Fossmark, R., et al., 2013. The effects of unilateral truncal vagotomy on gastric carcinogenesis in hypergastrinemic Japanese female cotton rats. Regul. Pept. 184, 62–67.

Gommerman, J.L., Rojas, O.L., Fritz, J.H., 2014. Re-thinking the functions of IgA(+) plasma cells. Gut Microbes 5 (5), 652–662.

He, C., Yang, Z., Lu, N., 2016. Imbalance of gastrointestinal microbiota in the pathogenesis of *Helicobacter pylori* -associated diseases. Helicobacter 21 (5), 337–348. Available from: http://doi.wiley.com/10.1111/hel.12297.

Heel, A.K., et al., 1997. Special article review: Peyer's patches. J. Gastroenterol. 12 (October 1996), 122–136.

Helander, H.F., Fandriks, L., 2014. Surface area of the digestive tract—revisited. Scand. J. Gastroenterol. 49 (6), 681–689.

Hoban, A.E., et al., 2016. Regulation of prefrontal cortex myelination by the microbiota. Transl. Psychiatry 6 (4), e774. Available from: http://www.ncbi.nlm.nih.gov/pubmed/27045844.

Hofstra, J.J., et al., 2015. Changes in microbiota during experimental human Rhinovirus infection. BMC Infect. Dis. 15, 336. Available from: http://www.ncbi.nlm.nih.gov/pubmed/26271750.

Horz, H.-P., 2015. Archaeal lineages within the human microbiome: absent, rare or elusive? Life 5 (2), 1333–1345. Available from: http://www.ncbi.nlm.nih.gov/pubmed/25950865.

Janson, L., Tischler, M., 2012. Medical Biochemistry: The Big Picture. McGraw-Hill Education, New York, NY.

Johansson, M.E.V., Hansson, G.C., 2016. Immunological aspects of intestinal mucus and mucins. Nat. Rev. Immunol. 16 (10), 639–649.

Jones, M., et al., 2006. Brain-gut connections in functional GI disorders: anatomic and physiologic relationships. Neurogastroenterol. Motil. 18 (2), 91–103. Available from: http://doi.wiley.com/10.1111/j.1365-2982.2005.00730.x.

Khoshdel, A., et al., 2013. Bifidobacterium longum NCC3001 inhibits AH neuron excitability. Neurogastroenterol. Motil. 25 (7), e478–e484. Available from: http://www.ncbi.nlm.nih.gov/pubmed/23663494.

Kilian, M., et al., 2016. The oral microbiome—an update for oral healthcare professionals. Br. Dent. J. 221 (10), 657–666. Available from: http://www.nature.com/doifinder/10.1038/sj.bdj.2016.865.

Kunze, W.A., et al., 2009. *Lactobacillus reuteri* enhances excitability of colonic AH neurons by inhibiting calcium-dependent potassium channel opening. J. Cell. Mol. Med. 13 (8b), 2261–2270. Available from: http://doi.wiley.com/10.1111/j.1582-4934.2009.00686.x.

Lee, S.H., 2015. Intestinal permeability regulation by tight junction: implication on inflammatory bowel diseases. Intestinal Res. 13 (1), 11.

Li, H., et al., 2015. The outer mucus layer hosts a distinct intestinal microbial niche. Nat. Commun. 6, 8292. Available from: http://www.ncbi.nlm.nih.gov/pubmed/26392213.

Lloyd-Price, J., Abu-Ali, G., Huttenhower, C., 2016. The healthy human microbiome. Genome Med. 8 (51), 1–11.

Lozupone, C.A., et al., 2012. Diversity, stability and resilience of the human gut microbiota. Nature 489 (7415), 220–230. Available from: http://www.nature.com/doifinder/10.1038/nature11550.

Luczynski, P., et al., 2016. Growing up in a bubble: using germ-free animals to assess the influence of the gut microbiota on brain and behavior. Int. J. Neuropsychopharmacol. 19 (8), 1–17.

Mabbott, N.A., et al., 2013. Microfold (M) cells: important immunosurveillance posts in the intestinal epithelium. Mucosal Immunol. 6 (4), 666–677.

Macpherson, A.J., Heikenwalder, M., Ganal-Vonarburg, S.C., 2016. The liver at the nexus of host-microbial interactions. Cell Host Microbe 20 (5), 561–571.

Maldonado-Contreras, A., et al., 2011. Structure of the human gastric bacterial community in relation to Helicobacter pylori status. ISME J. 5 (4), 574–579. Available from: http://www.nature.com/doifinder/10.1038/ismej.2010.149.

McVey Neufeld, K.A., et al., 2013. The microbiome is essential for normal gut intrinsic primary afferent neuron excitability in the mouse. Neurogastroenterol. Motil. 25 (2), 183–e88. Available from: http://www.ncbi.nlm.nih.gov/pubmed/23181420.

Methé, B.A., et al., 2012. A framework for human microbiome research. Nature 486 (7402), 215–221. Available from: http://www.nature.com/doifinder/10.1038/nature11209.

Mora, J.R., et al., 2006. Generation of gut-homing IgA-secreting B cells by intestinal dendritic cells. Science (New York, N.Y.) 314, 1157–1160.

Moran, G.W., et al., 2008. Enteroendocrine cells: neglected players in gastrointestinal disorders? Ther. Adv. Gastroenterol. 1 (1), 51–60.

Nardone, G., Compare, D., 2015. The human gastric microbiota: is it time to rethink the pathogenesis of stomach diseases? United European Gastroenterol J. 3 (3), 255–260. Available from: http://www.ncbi.nlm.nih.gov/pubmed/26137299.

National Institutes of Health, 2012. NIH Human Microbiome Project defines normal bacterial makeup of the body. Available from: https://www.nih.gov/news-events/news-releases/nih-human-microbiome-project-defines-normal-bacterial-makeup-body.

Nguyen, T.L.A., et al., 2015. How informative is the mouse for human gut microbiota research? Dis. Model. Mech. 8 (1), 1–16.

Norman, J.M., Handley, S.A., Virgin, H.W., 2014. Kingdom-agnostic metagenomics and the importance of complete characterization of enteric microbial communities. Gastroenterology 146 (6), 1459–1469. Available from: http://www.ncbi.nlm.nih.gov/pubmed/24508599.

Pabst, O., Cerovic, V., Hornef, M., 2016. Secretory IgA in the coordination of establishment and maintenance of the microbiota. Trends Immunol. 37 (5), 287–296.

Pei, Z., et al., 2004. Bacterial biota in the human distal esophagus. Proc. Natl. Acad. Sci. U. S. A. 101 (12), 4250–4255. Available from: http://www.ncbi.nlm.nih.gov/pubmed/15016918.

Perez-Burgos, A., et al., 2013. Psychoactive bacteria Lactobacillus rhamnosus (JB-1) elicits rapid frequency facilitation in vagal afferents. Am. J. Physiol. Gastrointest. Liver Physiol. 304 (2), G211–G220.

Perez-Burgos, A., et al., 2014. The gut-brain axis rewired: adding a functional vagal nicotinic "sensory synapse". FASEB J. 28 (7), 3064–3074. Available from: http://www.ncbi.nlm.nih.gov/pubmed/24719355.

Qin, J., et al., 2010. A human gut microbial gene catalogue established by metagenomic sequencing. Nature 464 (7285), 59–65. Available from: http://www.nature.com/doifinder/10.1038/nature08821.

Reboldi, A., Cyster, J.G., 2016. Peyer's patches: organizing B-cell responses at the intestinal frontier. Immunol. Rev. 271 (1), 230–245.

Reinoso Webb, C., et al., 2016. Protective and pro-inflammatory roles of intestinal bacteria. Pathophysiology 23 (2), 67–80. Available from: http://linkinghub.elsevier.com/retrieve/pii/S0928468016300025.

Rogers, M.B., et al., 2016. Acute appendicitis in children is associated with a local expansion of fusobacteria. Clin. Infect. Dis. 63 (1), 71–78. Available from: http://www.ncbi.nlm.nih.gov/pubmed/27056397.

Samuel, B.S., et al., 2007. Genomic and metabolic adaptations of Methanobrevibacter smithii to the human gut. Proc. Natl. Acad. Sci. 104 (25), 10643–10648. Available from: http://www.ncbi.nlm.nih.gov/pubmed/17563350.

Schulz, C., et al., 2016. The active bacterial assemblages of the upper GI tract in individuals with and without Helicobacter infection. Gut, p.gutjnl-2016-312904. Available from: http://www.ncbi.nlm.nih.gov/pubmed/27920199.

Segata, N., et al., 2012. Composition of the adult digestive tract bacterial microbiome based on seven mouth surfaces, tonsils, throat and stool samples. Genome Biol. 13 (6), R42. Available from: http://www.ncbi.nlm.nih.gov/pubmed/22698087.

Smith, H.F., et al., 2017. Morphological evolution of the mammalian cecum and cecal appendix. C.R. Palevol 16 (1), 39–57.

Snider, E.J., Freedberg, D.E., Abrams, J.A., 2016. Potential role of the microbiome in barrett's esophagus and esophageal adenocarcinoma. Dig. Dis. Sci. 61 (8), 2217–2225. Available from: http://link.springer.com/10.1007/s10620-016-4155-9.

Strugnell, R.A., Wijburg, O.L.C., 2010. The role of secretory antibodies in infection immunity. Nat. Rev. Microbiol. 8 (1), 656–667.

Thaiss, C.A., et al., 2016. The microbiome and innate immunity. Nature 535 (7610), 65–74.

Tortorella, C., Neri, G., Nussdorfer, G.G., 2007. Galanin in the regulation of the hypothalamic-pituitary-adrenal axis (Review). Int. J. Mol. Med. 19 (4), 639–647. Available from: http://www.ncbi.nlm.nih.gov/pubmed/17334639.

Track, N.S., 1980. The gastrointestinal endocrine system. Can. Med. Assoc. J. 122 (3), 287–292. Available from: http://www.ncbi.nlm.nih.gov/pubmed/6989456.

Trasande, L., et al., 2013. Infant antibiotic exposures and early-life body mass. Int. J. Obes. 37 (1), 16–23. Available from: http://www.ncbi.nlm.nih.gov/pubmed/22907693.

Underhill, D.M., Iliev, I.D., 2014. The mycobiota: interactions between commensal fungi and the host immune system. Nat. Rev. Immunol. 14 (6), 405–416. Available from: http://www.ncbi.nlm.nih.gov/pubmed/24854590.

Uribe, A., et al., 1994. Microflora modulates endocrine cells in the gastrointestinal mucosa of the rat. Gastroenterology 107 (5), 1259–1269.

van 't Hof, W., et al., 2014. Antimicrobial defense systems in saliva. Monogr. Oral Sci. 24, 40–51. Available from: http://www.ncbi.nlm.nih.gov/pubmed/24862593.

Wiles, T.J., et al., 2016. Host gut motility promotes competitive exclusion within a model intestinal microbiota. PLoS Biol. 14 (7), e1002517. Available from: http://www.ncbi.nlm.nih.gov/pubmed/27458727.

Yano, J.M., et al., 2015. Indigenous bacteria from the gut microbiota regulate host serotonin biosynthesis. Cell 161 (2), 264–276. Available from: http://linkinghub.elsevier.com/retrieve/pii/S0092867415002482.

Yasuda, K., et al., 2015. Biogeography of the intestinal mucosal and lumenal microbiome in the rhesus macaque. Cell Host Microbe 17 (3), 385–391. Available from: http://www.ncbi.nlm.nih.gov/pubmed/25732063.

Zhou, L., Foster, J.A., 2015. Psychobiotics and the gut-brain axis: in the pursuit of happiness. Neuropsychiatr. Dis. Treat. 11, 715–723. Available from: http://www.ncbi.nlm.nih.gov/pubmed/25834446.

CHAPTER 3

Gut Microbiota Throughout the Lifespan

Objectives

- To become familiar with the changes in gut microbiota composition that occur during fetal development, infancy (including preterm birth), childhood, adulthood, and older adulthood.
- To understand the factors that potentially affect gut microbiota development in early life: in particular, infant diet and the transition to solid foods.
- To learn about the health outcomes linked with gut microbiome composition at each stage of life.
- To become aware of preliminary research on gut eukaryotes and viruses.

CHANGES IN GUT BACTERIAL COMMUNITIES OVER A LIFETIME

The acquisition of the human gut microbiota is a complex process and is influenced by a number of genetic and environmental factors over the lifespan. The gut microbiome is dynamic in the first weeks of life, with lower diversity and higher variability; a transition to higher diversity and lower variability occurs as development proceeds. The gut microbiota reaches a relatively stable state in adults, although infection, antibiotics, and drastic changes in diet can lead to disruptions (David et al., 2014). In older adulthood, the composition of the gut microbiota begins to shift toward less diversity and more pro-inflammatory species. See Fig. 3.1 for a summary of the major changes that occur at different life stages.

The gut microbiome has an impact on numerous functions important to health, including digestion, nutrient acquisition, and modulation of the immune system, brain development, and even behavior (Bäckhed et al., 2012; Hansen et al., 2012). As such, the composition of the gut microbiota in early life is emerging as a factor in helping achieve and maintain good health in the years to come.

Gut Microbiota
https://doi.org/10.1016/B978-0-12-810541-2.00003-8

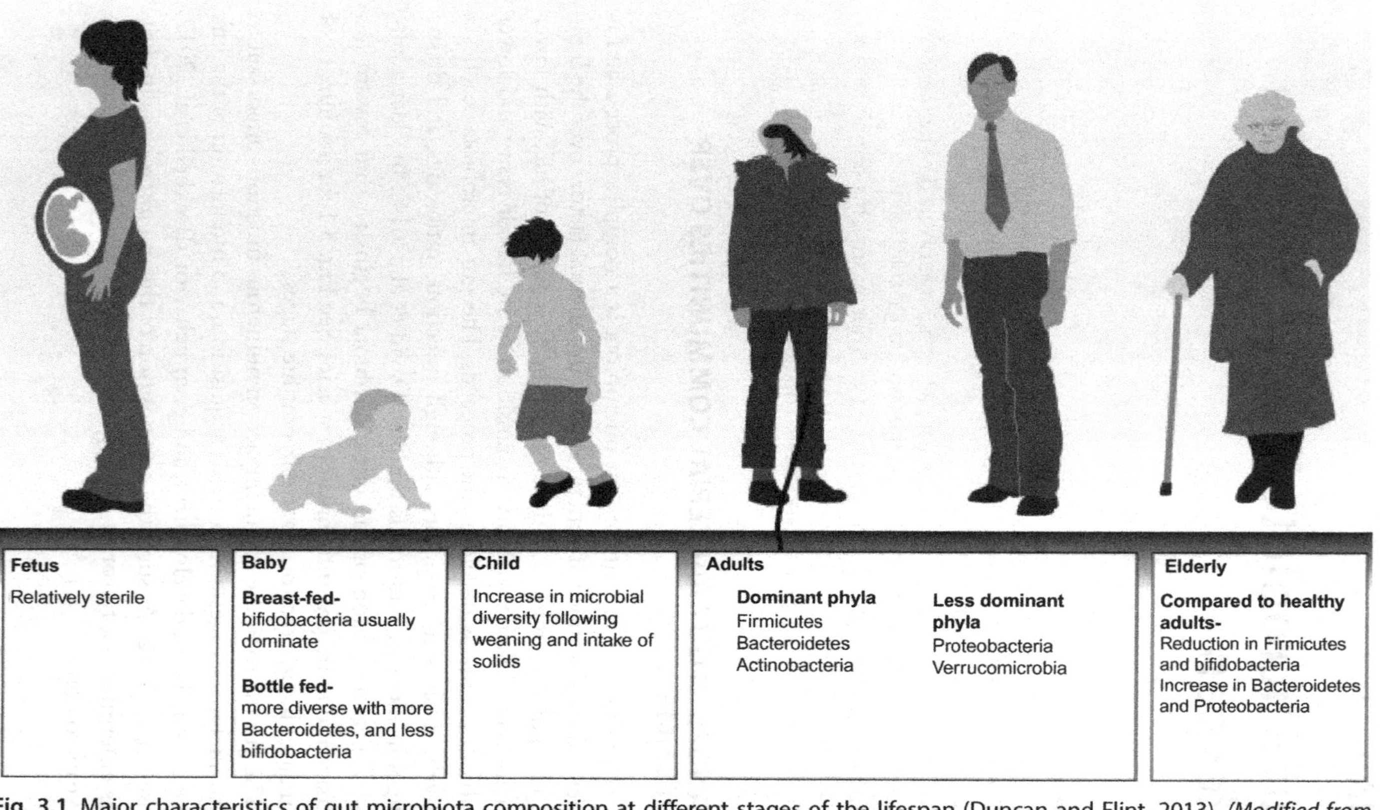

Fig. 3.1 Major characteristics of gut microbiota composition at different stages of the lifespan (Duncan and Flint, 2013). *(Modified from Duncan, S.H., Flint, H.J., 2013. Probiotics and prebiotics and health in ageing populations. Maturitas 75, 44–50.)*

First Exposures in the Intrauterine Environment

Until recently, scientists believed the intrauterine environment of a healthy pregnant woman was sterile, with the first colonization of the infant taking place at birth. The concept of the placenta being completely sterile is now being challenged. Bacterial DNA in the amniotic fluid, umbilical cord blood, meconium, and placental and fetal membranes from healthy pregnancies—independent of mode of delivery—has been discovered in the absence of any indication of infection or inflammation (Jimenez et al., 2005; Satokari et al., 2009). The placenta is now described as harboring a set of culturable bacteria. Thus, the neonate may be exposed to bacteria and/or bacterial products much earlier than initially believed (Aagaard et al., 2014). Prior to these discoveries, the presence of bacteria in clinical cultures—in particular, Gram-negative bacteria in amniotic fluid—was diagnostic for adverse pregnancy outcomes such as miscarriage, preterm delivery, and premature rupture of membranes (Bearfield et al., 2002). New research has found that the placentas of subjects with a history of antenatal infection and antibiotic treatment and of those who experienced preterm birth have different groupings of bacterial taxa as compared to those of women with healthy pregnancies (Aagaard et al., 2014). Thus, the type of bacterial taxa present, not the mere occurrence of bacteria in the placenta, may be what initiates intrauterine infection and leads to adverse pregnancy outcomes.

A mother-to-child transfer of commensal bacteria through the placental barrier may occur, but infant colonization remains unclear. The placental bacterial community appears to be largely composed of nonpathogenic commensal microbiota from the phyla Firmicutes, Tenericutes, Proteobacteria, Bacteroidetes, and Fusobacteria; however, contamination cannot be ruled out in many studies. Remarkably, the placental microbes do not appear to closely resemble those in the mother's stool or vaginal microbiomes, but closely resemble her oral microbiome in the regions of the supragingival plaque (i.e., the bacterial plaque located above the gumline) and the dorsum of the tongue (Aagaard et al., 2014). This implies that the bulk of placental bacteria are likely not contaminants of the stool or the vagina. Instead, they could be seeded mainly from the oral cavity. Further study in this area, however, is needed to confirm where intrauterine bacteria originate from and how they impact the neonate's developing immune system and general health.

The Infant Microbiome

Despite some possible bacterial exposures before birth, the fetus is still relatively sterile until it encounters the environment outside its mother. Birth represents an opportunity for exposure to a wide variety of microbes in the environment, including the maternal microbiota. The act of passing through the birth canal in a vaginal delivery and thus coming into contact with the maternal feces and vaginal microbiota influences the infant's gut microbiota composition at birth. The infant born vaginally acquires bacterial communities resembling its own mother's vaginal and fecal microbiota (Dominguez-Bello et al., 2010). The first colonizers of the infant gut in vaginal birth are facultative anaerobic bacteria, such as *Staphylococcus*, *Streptococcus*, *Enterococcus*, and *Enterobacter* species, followed by anaerobes such as *Bifidobacterium*, *Bacteroides*, and *Clostridium* species (Martin et al., 2016).

Surgical delivery by Cesarean section (C-section) is associated with altered colonization of the infant gut at birth, ostensibly because the infant is not exposed to the maternal microbiota in the same manner as in vaginal birth. In contrast with those born vaginally, infants born by C-section harbor bacterial communities similar to those found on the skin surface and in the hospital environment (e.g., health-care workers, hospital surfaces, and other newborns) (Dominguez-Bello et al., 2010). The gut microbiota of infants delivered by C-section appears to be less diverse, in terms of bacterial species, than the microbiota of vaginally delivered infants (Biasucci et al., 2008); it is characterized by a substantial absence of *Bifidobacterium* and *Bacteroides* species, accompanied by an increase in the presence of *Clostridium difficile* (Penders et al., 2006; Biasucci et al., 2008). Colonization of the infant gut is also affected by elective versus emergency Cesarean delivery, as bacterial richness and diversity is lowest among infants born by elective C-section and highest among those born by emergency C-section (Azad et al., 2013). Researchers speculate that during emergency C-section, the infant is still exposed to many species from its mother's microbiome before surgery is initiated, which may account for the differences.

A 2016 pilot study (Dominguez-Bello et al., 2016) documented a "microbial restoration procedure" in which four infants delivered by C-section were swabbed with maternal vaginal fluids at birth. In the procedure, sterile gauze was incubated in the vagina of the mother prior to the surgery and was used to swab the baby's mouth, face, and body within the first two minutes after C-section birth. The gut, mouth, and skin bacterial communities of the infants who received the intervention were enriched in vaginal bacteria that were underrepresented in C-section infants who had not received the intervention. The health risks and benefits of this intervention, however, remain unknown; it could potentially lead to health outcomes completely different from either vaginal birth or C-section birth, since the intervention yielded a unique microbiota composition that did not match either group.

Regardless of delivery mode, the period after birth is a time when environmental microbes and oral and skin microbes from the mother and other caregivers are transferred to the infant via feeding, kissing, and caressing. Furthermore, the infant is continuously exposed to new microbes through its early diet.

Links to Health

One of the most important triggers for immune system development is the exposure to microbial components immediately after birth, leading researchers to wonder whether varying microbial exposures that accompany different modes of delivery could have lasting effects on health. An increasing body of evidence suggests that children born by C-section are at increased risk of allergies, such as allergic rhinoconjunctivitis, and asthma (Renz-Polster et al., 2005; Roduit et al., 2009) compared with those born vaginally. In addition, Cesarean delivery is associated with increased body mass and childhood obesity (Blustein et al., 2013). So far, causal links between these conditions and altered microbial colonization in C-section delivery have not been proven.

Recent work has suggested the observed differences in gut microbiota composition between infants born vaginally and those born by C-section may in fact be temporary, calling into question the impact of these changes on future health. A study that explored both composition and function of the microbiota at different body sites (the stool, oral gingiva, nares, skin, and vagina) of mother-infant dyads, from pregnancy to 6 weeks postdelivery, found the neonatal microbiota and its associated functional pathways were relatively similar across all body sites at the time of delivery, except in the meconium. While immediately after birth certain body sites (the oral gingiva, nares, and skin) of infants born by C-section showed minor differences in composition, delivery mode was not associated with differences in bacterial community function. By 6 weeks after delivery, however, microbiota structure and function had expanded, and there were no differences between infants delivered vaginally and those delivered by Cesarean section. Rather, body site was the strongest determinant of bacterial community composition and function (Chu et al., 2017).

Could gut microbiota still account for the known associations between C-section and later disease, even if birth by C-section does not significantly alter the infant microbiota? Researchers from the same group at Baylor College of Medicine (the United States) point out that several factors known to alter the gut microbiome are associated with a higher rate of Cesarean surgery: diet, antibiotic exposure, gestational age, and host genetics (Aagaard et al., 2016). They say that these factors, which tend to co-occur with C-section, could potentially drive gut microbiota alterations that might turn out to have a causal relationship with health outcomes. The researchers highlight diet as an important factor: a human study from their group showed a maternal high-fat diet during gestation and lactation was associated with a distinct microbiome in the newborn's stool (independent of maternal body mass index), with a notable depletion of *Bacteroides* species in these neonates, which persisted through 6 weeks of age (Chu et al., 2016). This led the authors to hypothesize that maternal diet, through its effects on the placental microbiota, could drive changes in infant microbiota composition after birth with possible health consequences for the child.

While it remains clear that birth represents an important time of microbial colonization for the infant, future work will identify how gestational and birth factors work together to shape the infant microbiota and link to health outcomes later in life.

The Preterm Infant

The infant born before 37 weeks' gestation faces significant challenges because of organ immaturity, in addition to frequent antibiotic exposure and an extended period in the hospital neonatal unit. Not surprisingly, the gut microbiota of a preterm infant differs from that of a full-term infant (Arboleya et al., 2012; Grześkowiak et al., 2015).

The preterm infant displays altered intestinal colonization by commensal microorganisms, increased occurrence of potential pathogens, and high interindividual variability and reduced microbial diversity when compared with a healthy term infant (Arboleya et al., 2012). Studies indicate that preterm infants harbor increased levels of facultative anaerobic microorganisms, such as Enterobacteriaceae, Enterococcaceae, and *Lactobacillus* spp., together with reduced levels of anaerobes, including *Bifidobacterium* (for instance, *B. longum*), *Bacteroides*, and *Atopobium* spp. (Arboleya et al., 2012; Grześkowiak et al., 2015).

Links to Health

Alterations in the microbiota of preterm infants may lead to delayed maturation of the immune system, which could have profound effects on the health of the infant; in particular, it may increase the risk of infection. Given that dominance of bifidobacteria, especially *B. longum* and *B. lactis*, appear important for normal gut microbiota development during the first weeks of life, further research will determine whether the reduced levels of *Bifidobacterium* species in preterm infants have implications for future health.

Necrotizing enterocolitis (NEC) is a serious and sometimes fatal gastrointestinal disease occurring in preterm infants. While links between intestinal bacterial colonization and NEC were queried several decades ago (Sántulli et al., 1975), no single bacterium could be identified as a trigger. More recent work, however, suggests that disruptions in a premature infant's gut microbiota precede the onset of NEC: namely, an increase in Proteobacteria and a decrease in Firmicutes occur before NEC diagnosis (Mai et al., 2011). Although it is still unclear whether gut microbiota changes represent a cause or an effect of NEC, researchers currently hypothesize that immune abnormalities render the GI tracts of premature infants "hyperreactive" as compared with those of their full-term peers and that in the presence of certain host genes, the microbes colonizing the GI tract can promote NEC development (Niño et al., 2016).

Impact of Early Diet

Mode of feeding—breast milk versus formula—has a significant impact on the composition of the microbiota in early infancy. Exclusive breastfeeding in the first 3 months of life has long-lasting effects on microbial colonization (Martin et al., 2016). The gut microbiota of breastfed infants is dominated by Actinobacteria: in particular, beneficial species of *Bifidobacterium* and *Lactobacillus* (Harmsen et al., 2000). Bifidobacteria in the breastfed infant rapidly colonize the gut and remain until weaning. The formula-fed infant develops a more diverse microbiota, with fewer bifidobacteria and more pathogens like *Clostridium coccoides*, *Staphylococcus* spp., and those from the family Enterobacteriaceae (Fallani et al., 2010; Harmsen et al., 2000).

Human breast milk contains, in addition to live bacteria, a great variety of complex carbohydrates called **human milk oligosaccharides** (HMOs) as illustrated in Fig. 3.2 (Zivkovic et al., 2011). HMOs are not digestible by infants. Instead, these sugars are digested by selected bacteria in the infant GI tract that are genetically equipped to break them down: specifically *Bifidobacterium infantis* (Ward et al., 2006). HMOs therefore modulate the intestinal microbiota of breastfed infants and act as prebiotics by enriching certain beneficial bacteria and promoting the release of short-chain fatty acids that feed the infant gut.

Links to Health

Epidemiological data show that breastfeeding is associated not only with short-term benefits—namely, a lowered risk of infectious diseases in

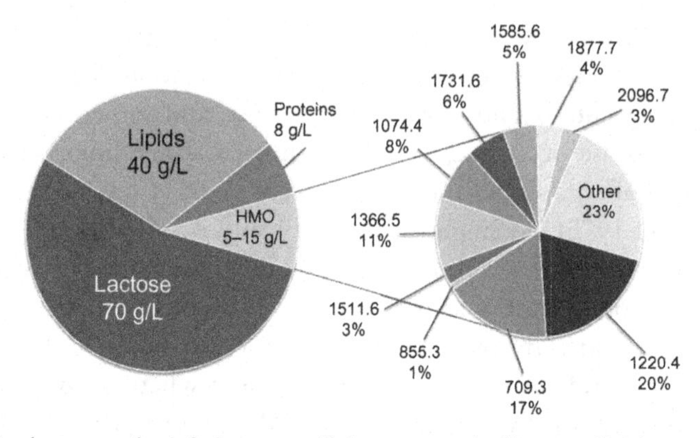

Fig. 3.2 As shown on the left, human milk is composed of lactose, lipids, proteins, and human milk oligosaccharides (HMOs). The HMOs are the third most abundant component of breast milk, at 5–15 g/L (grams per liter). The pie chart on the right shows a breakdown of the HMOs that are the most abundant in pooled human milk samples. Mass spectrometry can be used to identify specific oligosaccharide molecules through accurate mass measurements; individual HMO structures are labeled in the chart with their masses and relative abundance (%). *(From Zivkovic, A. M. et al., 2011. Human milk glycobiome and its impact on the infant gastrointestinal microbiota. Proc. Natl. Acad. Sci. U. S. A. 108.)*

childhood—but also with longer-term benefits: a lower prevalence of overweight/obesity, protection against type 2 diabetes, and even increased performance on intelligence tests (Horta and Victora, 2013). To date, a mechanistic link has not been shown between these health effects and early-life changes in gut microbiota driven by breastfeeding.

Impact of the Introduction of Solid Foods

Weaning is the most significant factor shifting the microbiota toward a more diverse and stable adult-like composition (Fallani et al., 2010). Gut microbiota composition appears to be strongly affected by the transition from breastfeeding to "family foods" (solids) high in protein and fiber, with a greater microbiota diversity associated with infant intake of meats, cheeses, and high-fiber bread (Laursen et al., 2016). Following the introduction of solid foods, the early colonizers of the gut are replaced with a more complex microbiota (Martin et al., 2016). Bifidobacteria still dominate the intestinal microbiota, but their proportions significantly decrease as the microbial community begins to diversify (Fallani et al., 2011). Facultative anaerobes decrease, while proportions of strictly anaerobic clostridia increase. Proportions of *Bacteroides* remain unchanged and are one of the most predominant groups in the infant gut microbiota after weaning commences. In general, the introduction of solid foods is associated with a higher prevalence butyrate-producing bacteria like the *Clostridium coccoides* group (Martin et al., 2016); the increased prevalence of these bacteria following the introduction of solid foods might be explained by their ability to easily metabolize the complex carbohydrates that have been introduced into the diet.

Childhood and Adolescence

While initial studies suggested the gut microbiome, following early-life colonization, became adultlike by age 2, new evidence shows it continues to mature well beyond 2 years of age. Childhood appears to represent a unique transitional stage with respect to the gut microbiome. Although a health-associated pediatric gut microbiome has several adultlike features, it also retains many of its own distinct compositional and functional qualities (Hollister et al., 2015). Fig. 3.3 shows the general trend of gut microbiota development in the infant and child.

The normal pediatric gut microbiome is composed largely of bacteria belonging to Bacteroidetes and Firmicutes, and the ratio of these two phyla varies considerably across individuals (Hollister et al., 2015). In contrast with adults, the average healthy child has a gut community with a significantly lower abundance of Bacteroidetes and significantly greater abundances of Firmicutes and Actinobacteria (Ringel-Kulka et al., 2013; Hollister et al., 2015). Despite many taxa being shared between children and adults, adults harbor a greater proportion of *Bacteroides* spp. and a lower abundance of *Bifidobacterium* spp. (Ringel-Kulka et al., 2013). Overall, a major difference between the gut microbiota of adults as compared with children is the microbial diversity—that is, with age comes increased diversity

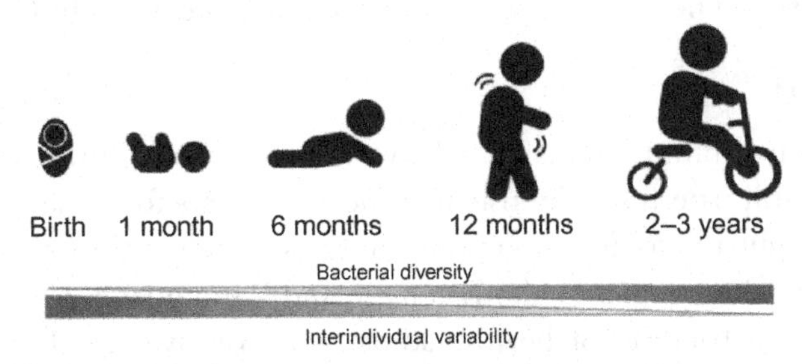

Fig. 3.3 Microbiota composition of the infant and child intestine changes through the first several years of life: bacterial diversity gradually increases, while interindividual variability gradually decreases (Arrieta et al., 2014). *(Modified from Arrieta, M.-C., et al., 2014. The intestinal microbiome in early life: health and disease. Front. Immunol. 5, 427. Available at http://www.ncbi.nlm.nih.gov/pubmed/25250028.)*

(Hollister et al., 2015; Ringel-Kulka et al., 2013). Although the reasons for lower microbial diversity in children are not clear, they are likely related to children's more limited environmental and dietary exposures.

A healthy child's gut microbiome is also unique from a functional perspective. Compared with adults, children have gut microbial communities that are enriched in functions that may support ongoing physical development (Hollister et al., 2015). On the other hand, the microbiota of adults is enriched in functions associated with inflammation and increased risk of adiposity (or obesity). The most notable differences include the enrichment of genes involved in the syntheses of vitamin B_{12} and folate. In the case of folate, children's microbial communities are enriched with genes that support DNA synthesis, replication, and repair, which are necessary for growth and development; adult microbial communities are enriched in genes that support dietary utilization of folate. Furthermore, the microbiota produces vitamin B_{12} (cobalamin), which has antiinflammatory and antioxidant benefits and is essential for neurological function. Blood concentrations of cobalamin peak around the age of 7, suggesting that the microbiota later decrease production as they adapt to support the needs of the adult.

Research has also confirmed that the gut microbiome of adolescents (aged 11–18 years) differs from that of adults (Agans et al., 2011), in particular at the genus level. Most strikingly, a higher prevalence of *Bifidobacterium* spp. (close to a twofold difference) is also seen in adolescents compared with adults.

The intestinal microbiota of children and adolescents has unique characteristics and continues to mature well beyond infancy, suggesting that it may still be vulnerable to external exposures. Besides early diet, other factors that may influence the early colonization process include antibiotic exposure, farm exposure, place of birth, and the presence of siblings and household pets. These factors will be covered in more detail in Chapter 5.

Adulthood

The gut microbiome remains relatively stable throughout adulthood. While it is still true that scientists have no fixed rules for what constitutes a "normal" gut microbiota composition, some patterns do exist. Bacteria belonging to the families Lachnospiraceae and Ruminococcaceae codominate the fecal bacteria of healthy adults with, on average, 10%–45% of the total fecal bacteria belonging to Lachnospiraceae and 16%–27% to Ruminococcaceae. Members of the Bacteroidaceae/Prevotellaceae family comprise the remaining 12%–60%. At the phylum level, Firmicutes (families Lachnospiraceae and Ruminococcaceae), Bacteroidetes (Bacteroidaceae,

Prevotellaceae, and Rikenellaceae), and Actinobacteria (Bifidobacteriaceae and Coriobacteriaceae) represent the majority of bacteria (Maukonen and Saarela, 2015). *Bifidobacterium* species, *Lactobacillus* group, and bacteria within the family Coriobacteriaceae are much lower in adulthood than in early childhood.

When an individual's gut microbiota has reached its stable peak community, the composition appears to persist for a significant period of time. For example, a sample of 37 healthy adults found that 60% of the original strains were present 5 years later (Faith et al., 2013). The main external factors that can affect the composition of the microbiota in healthy adults (discussed in Chapter 5) include infection, major dietary changes, and antibiotic therapy or other medications.

Older Adulthood

The gut microbiota undergoes substantial changes with aging, shifting gradually over time. Although relatively few studies have been carried out in these populations, it is clear that microbial composition differences exist in older adults (Claesson et al., 2011, 2012; O'Toole and Claesson, 2010) when compared with other age groups. The gut microbiota of older adults displays greater interindividual variation than that of younger adults (Claesson et al., 2011). Aging is associated with a dramatic shift in composition of the core microbiota—in general, with increasing age comes a greater abundance of subdominant bacterial species. Increases in the relative abundance of Bacteroidetes (over Firmicutes) and Proteobacteria are observed (Enck et al., 2009; Odamaki et al., 2016), with reductions in Actinobacteria, including important *Bifidobacterium* strains, as aging proceeds.

The most profound changes in the gut microbiota seem to occur at the extreme end of the lifespan (Biagi et al., 2010): remarkably, the microbiota of a centenarian differs from that of 70-year-old. The centenarian's microbiota is characterized by an increase in pathobionts (e.g., species of *Fusobacterium*, *Bacillus*, *Staphylococcus*, *Corynebacterium*, and those from the family Micrococcaceae) with a decrease in the numbers of butyrate-producing bacteria (*Faecalibacterium prausnitzii*, *Eubacterium rectale*, *E. hallii*, and *E. ventriosum*).

The ELDERMET consortium was formed in 2007 and examined the associations between the gut microbiota, diet, and health in 500 healthy elderly volunteers. This work supported previous findings that the constituents of the gut microbiota in older people (>65 years) are extremely variable from individual to individual and that the core microbiota and diversity levels differ from those of younger adults (Claesson et al., 2012). Data from this project provided evidence that aging *per se* may have less influence on gut microbiota composition than the lifestyle changes that often occur with age. Residence of an older adult (i.e., community, day hospital, rehabilitation facility, or long-term care) has an impact on overall microbial diversity: the gut microbiota of older adults in long-term care is significantly less diverse than that of community dwellers. Furthermore, the type of diet an older adult consumes is a key factor in shaping gut microbiota. An older adult living in a community setting, who consumes a low-fat/higher-fiber diet that includes a variety of foods, has a much greater microbial diversity compared with an older adult living in long-term care; researchers have observed that long-term care residents, who tend to consume a high-fat/low-fiber diet that lacks variety, have the least diverse gut microbiota.

Links to Health

The notable decrease in bifidobacterial strains observed in older adults may be a factor in this population's increased risk of pathogenic infection (Leung and Thuret, 2015). The proliferation of pathobionts at the cost of beneficial bacteria in older age is believed to contribute to a heightened inflammatory status and may be a risk factor for chronic health conditions (O'Toole and Claesson, 2010). Lower microbial diversity is associated with increased frailty (Jackson et al., 2016) and inflammation (Claesson et al., 2012). Species more abundant with frailty include *Eubacterium dolichum* and *Eggerthella lenta* (Jackson et al., 2016). While a causal relationship between gut microbiota and frailty has not been established, the association warrants attention in future studies since frailty is a better predictor of poor health outcomes than chronological age.

GUT EUKARYOTES AND VIRUSES

Relatively few studies have focused on gut eukaryotes and viruses throughout the lifespan. In the case of eukaryotes (e.g., fungi), researchers have little information about their contributions to gut activities and to health outcomes. Metagenomic analyses of human fecal samples have indicated that viruses of eukaryotes exist in the gut microbiota. In the limited studies that have been done, however, their role in human health is uncertain (Duerkop and Hooper, 2013; Lecuit and Eloit, 2013) except in cases of known viral pathogens for example, the recent discovery of transplacental infection of infants by Zika virus, resulting in microcephaly (Mysorekar et al., 2016;

Calvet et al., 2016). The majority of viruses in the gut are bacteriophages, but unlike the bacteria residing in the gut, the virome is mostly unexplored. Preliminary studies have noted a high degree of interpersonal variation in virome composition (Minot et al., 2011; Reyes et al., 2011, 2015; Lim et al., 2015). Furthermore, a genetic influence is suggested by studies on infant monozygotic twin pairs that indicate the viromes of these cotwins are more similar to each other than they are to the viromes of unrelated infants (Reyes et al., 2015; Lim et al., 2015). The viromes of adult cotwins, however, are substantially different from each other (Reyes et al., 2011). Thus, the available literature suggests that the viromes exhibit a high degree of interpersonal variability except in the cases of infant monozygotic twins. Further studies are necessary to determine the origin of gut viruses, the factors that determine gut virome composition throughout the lifespan, and the virome's possible effects on health.

REFERENCES

Aagaard, K., et al., 2014. The placenta harbors a unique microbiome. Sci. Transl. Med. 6 (237), 237ra65.

Aagaard, K., Stewart, C.J., Chu, D., 2016. Una destinatio, viae diversae. EMBO Rep. 17 (12), 1679–1684. Available from: http://embor.embopress.org/lookup/doi/10.15252/embr.201643483.

Agans, R., et al., 2011. Distal gut microbiota of adolescent children is different from that of adults. FEMS Microbiol. Ecol. 77 (2), 404–412.

Arboleya, S., et al., 2012. Establishment and development of intestinal microbiota in preterm neonates. FEMS Microbiol. Ecol. 79 (3), 763–772.

Arrieta, M.-C., et al., 2014. The intestinal microbiome in early life: health and disease. Front. Immunol. 5, 427. Available from: http://www.ncbi.nlm.nih.gov/pubmed/25250028.

Azad, M.B., et al., 2013. Gut microbiota of healthy Canadian infants: profiles by mode of delivery and infant diet at 4 months. CMAJ 185 (5), 385–394.

Bäckhed, F., et al., 2012. Defining a healthy human gut microbiome: current concepts, future directions, and clinical applications. Cell Host Microbe 12 (5), 611–622.

Bearfield, C., et al., 2002. Possible association between amniotic fluid micro-organism infection and microflora in the mouth. BJOG Int. J. Obstet. Gynaecol. 109 (5), 527–533.

Biagi, E., et al., 2010. Through ageing, and beyond: gut microbiota and inflammatory status in seniors and centenarians. PLoS One 5 (5).

Biasucci, G., et al., 2008. Cesarean delivery may affect the early biodiversity of intestinal bacteria. J. Nutr. 138 (9), 1796S–1800S.

Blustein, J., et al., 2013. Association of caesarean delivery with child adiposity from age 6 weeks to 15 years. Int. J. Obes. 37 (7), 900–906.

Calvet, G., et al., 2016. Detection and sequencing of Zika virus from amniotic fluid of fetuses with microcephaly in Brazil: a case study. Lancet Infect. Dis. 16 (6), 653–660.

Chu, D.M., et al., 2016. The early infant gut microbiome varies in association with a maternal high-fat diet. Genome Med. 8, 77.

Chu, D.M., et al., 2017. Maturation of the infant microbiome community structure and function across multiple body sites and in relation to mode of delivery. Nat. Med. Available from: http://www.nature.com/doifinder/10.1038/nm.4272.

Claesson, M.J., et al., 2011. Composition, variability, and temporal stability of the intestinal microbiota of the elderly. Proc. Natl. Acad. Sci. U. S. A. 108 (Suppl.), 4586–4591.
Claesson, M.J., et al., 2012. Gut microbiota composition correlates with diet and health in the elderly. Nature 488 (7410), 178–184.
David, L.A., et al., 2014. Diet rapidly and reproducibly alters the human gut microbiome. Nature 505 (7484), 559–563.
Dominguez-Bello, M.G., et al., 2010. Delivery mode shapes the acquisition and structure of the initial microbiota across multiple body habitats in newborns. Proc. Natl. Acad. Sci. U. S. A. 107 (26), 11971–11975.
Dominguez-Bello, M.G., et al., 2016. Partial restoration of the microbiota of cesarean-born infants via vaginal microbial transfer. Nat. Med. 22 (3), 250–253. Available from: http://www.ncbi.nlm.nih.gov/pubmed/26828196.
Duerkop, B.A., Hooper, L.V., 2013. Resident viruses and their interactions with the immune system. Nat. Immunol. 14 (7), 654–659.
Duncan, S.H., Flint, H.J., 2013. Probiotics and prebiotics and health in ageing populations. Maturitas 75, 44–50.
Enck, P., et al., 2009. The effects of ageing on the colonic bacterial microflora in adults. Z. Gastroenterol. 47 (7), 653–658.
Faith, J.J., et al., 2013. The long-term stability of the human gut microbiota. Science (New York, N.Y.) 341 (6141), 1237439.
Fallani, M., et al., 2010. Intestinal microbiota of 6-week-old infants across Europe: geographic influence beyond delivery mode, breast-feeding, and antibiotics. J. Pediatr. Gastroenterol. Nutr. 51 (1), 77–84.
Fallani, M., et al., 2011. Determinants of the human infant intestinal microbiota after the introduction of first complementary foods in infant samples from five European centres. Microbiology 157 (5), 1385–1392.
Grześkowiak, Ł., et al., 2015. Gut Bifidobacterium microbiota in one-month-old Brazilian newborns. Anaerobe 35, 54–58.
Hansen, C.H.F., et al., 2012. Patterns of early gut colonization shape future immune responses of the host. PLoS One 7 (3).
Harmsen, H.J., et al., 2000. Analysis of intestinal flora development in breast-fed and formula-fed infants by using molecular identification and detection methods. J. Pediatr. Gastroenterol. Nutr. 30 (1), 61–67.
Hollister, E.B., et al., 2015. Structure and function of the healthy pre-adolescent pediatric gut microbiome. Microbiome 3 (1), 36.
Horta, B.L., Victora, C.G., 2013. Long-term effects of breastfeeding … a systematic review …. Available from: www.who.int/about/licensing/copyright_form/en/index.html.
Jackson, M., et al., 2016. Signatures of early frailty in the gut microbiota. Genome Med. 8 (1), 8.
Jimenez, E., et al., 2005. Isolation of commensal bacteria from umbilical cord blood of healthy neonates born by cesarean section. Curr. Microbiol. 51 (4), 270–274.
Laursen, M.F., et al., 2016. Infant gut microbiota development is driven by transition to family foods independent of maternal obesity. mSphere 1 (1), 1–16.
Lecuit, M., Eloit, M., 2013. The human virome: new tools and concepts. Trends Microbiol. 21 (10), 510–515.
Leung, K., Thuret, S., 2015. Gut microbiota: a modulator of brain plasticity and cognitive function in ageing. Healthcare 3 (4), 898–916. Available from: http://www.mdpi.com/2227-9032/3/4/898/.
Lim, E.S., et al., 2015. Early life dynamics of the human gut virome and bacterial microbiome in infants. Nat. Med. 21 (10), 1228–1234.
Mai, V., et al., 2011. Fecal microbiota in premature infants prior to necrotizing enterocolitis. In: Chakravortty, D. (Ed.), PLoS One 6 (6), e20647. Available from: http://dx.plos.org/10.1371/journal.pone.0020647.

Martin, R., et al., 2016. Early-life events, including mode of delivery and type of feeding, siblings and gender, shape the developing gut microbiota. PLoS One 11 (6), e0158498.
Maukonen, J., Saarela, M., 2015. Human gut microbiota: does diet matter? Proc. Nutr. Soc. 74 (1), 23–36.
Minot, S., et al., 2011. The human gut virome: inter-individual variation and dynamic response to diet. Genome Res. 21 (10), 1616–1625.
Mysorekar, I.U., et al., 2016. Modeling zika virus infection in pregnancy. N. Engl. J. Med. 375 (5), 481–484.
Niño, D.F., Sodhi, C.P., Hackam, D.J., 2016. Necrotizing enterocolitis: new insights into pathogenesis and mechanisms. Nat. Rev. Gastroenterol. Hepatol. 13 (10), 590–600. Available from: http://www.ncbi.nlm.nih.gov/pubmed/27534694.
Odamaki, T., et al., 2016. Age-related changes in gut microbiota composition from newborn to centenarian: a cross-sectional study. BMC Microbiol. 16 (1), 90.
O'Toole, P.W., Claesson, M.J., 2010. Gut microbiota: changes throughout the lifespan from infancy to elderly. Int. Dairy J. 20 (4), 281–291.
Penders, J., et al., 2006. Factors influencing the composition of the intestinal microbiota in early infancy. Pediatrics 118 (2), 511–521.
Renz-Polster, H., et al., 2005. Caesarean section delivery and the risk of allergic disorders in childhood. Clin. Exp. Allergy 35 (11), 1466–1472.
Reyes, A., et al., 2011. Viruses in the faecal microbiota of monozygotic twins and their mothers. Nature 466 (7304), 334–338.
Reyes, A., et al., 2015. Gut DNA viromes of Malawian twins discordant for severe acute malnutrition. Proc. Natl. Acad. Sci. U. S. A. 112 (38), 11941–11946.
Ringel-Kulka, T., et al., 2013. Intestinal microbiota in healthy U.S. young children and adults-a high throughput microarray analysis. PLoS One 8 (5).
Roduit, C., et al., 2009. Asthma at 8 years of age in children born by caesarean section. Thorax 64 (2), 107–113.
Sántulli, T.V., et al., 1975. Acute necrotizing enterocolitis in infancy: a review of 64 cases. Pediatrics 55 (3), 376–387. Available from: http://www.ncbi.nlm.nih.gov/pubmed/1143976.
Satokari, R., et al., 2009. Bifidobacterium and Lactobacillus DNA in the human placenta. Lett. Appl. Microbiol. 48 (1), 8–12.
Ward, R.E., et al., 2006. In vitro fermentation of breast milk oligosaccharides by Bifidobacterium infantis and Lactobacillus gasseri. Appl. Environ. Microbiol. 72 (6), 4497–4499. Available from: http://www.ncbi.nlm.nih.gov/pubmed/16751577.
Zivkovic, A.M., et al., 2011. Human milk glycobiome and its impact on the infant gastrointestinal microbiota. Proc. Natl. Acad. Sci. U. S. A. 108, 4653–4658.

CHAPTER 4

Gut Microbiota in Health and Disease

Objectives

- To become familiar with how the immune system normally maintains homeostasis in the gut.
- To learn about gut microbiota and immune system activity in relation to (1) diseases caused by important microbial enteric pathogens and (2) various complex diseases (including some brain-related conditions).
- To understand what is known about features of "health-associated" and "disease-associated" microbiomes.

Many microorganisms have a well-known role in causing disease. Since the germ theory of disease was advanced (as explained in Chapter 1), a set of criteria called **Koch's postulates**, published in 1890, have been used to identify the specific microbial pathogen that causes a particular disease. Fulfilling Koch's postulates requires the isolation of the putative pathogen from infected tissue and the subsequent demonstration that this isolate will cause disease when it is inoculated into a healthy subject. Finding the pathogens responsible for conditions like cholera and tuberculosis that once caused millions of deaths has led not only to an improved understanding of these diseases, but also to the development of strategies for treatment and prevention—one of the biggest advances in the history of medicine.

The new layer of knowledge about the normal gut microbiota that has emerged in the last several decades adds a further layer of complexity to this pathogen model. Beyond single pathogens causing disease, it is now considered possible that alterations in the structure or function of the gut microbiota—involving multiple species and relationships—play a causal role in the pathogenesis or maintenance of many diseases. While no such cases are definitively proved to date, the evidence detailed below demonstrates that the immune system and its relationship with microbes could be at the core of various chronic conditions.

Gut Microbiota
https://doi.org/10.1016/B978-0-12-810541-2.00004-X

MICROBIAL TOLERANCE AND MAINTENANCE OF GUT HOMEOSTASIS

In the preceding chapters, the gut has been described as a reservoir of many foreign materials ranging from gut microorganisms to food particles as well as other substances from the environment. Some of these foreign materials could potentially harm the body, so the gut's immune tissue (gut-associated lymphoid tissue or GALT as described in Chapter 2) is charged with maintaining homeostasis within this environment. Notably, GALT is part of what is best described as the "common mucosal immune system," which also includes immune tissues of the respiratory tract and the genitourinary tract. Immune cells activated in any specific region of this common immune system may enter the circulation and hone in on any of the other mucosal tissues; thus, the different mucosal immune systems are interconnected.

The systemic (nonmucosal) immune system responds to foreign materials by activating a protective immune defense against them. In contrast, GALT normally functions to establish and maintain a homeostasis with foreign materials in the gut—unless a disruption leads to intestinal inflammation and the induction of a protective immune mechanism. GALT must normally maintain a delicate balance between inducing host tolerance to its foreign commensal gut microbiota and, at the same time, providing protection against infections by gut pathogens (Swiatczak and Cohen, 2015). To achieve these objectives, direct contact between the gut contents and the epithelial surface is minimized (Hooper and Macpherson, 2010). The combined effects of physical barriers (discussed in Chapter 2) such as epithelial tight junctions and mucus secretions, and chemical barriers such as antimicrobial peptides and secretory immunoglobulin A (sIgA) reduce the probability of microorganisms or food components translocating into the lamina propria. The result is a minimization of pathogen invasion and a sequestering of commensal microbes to prevent the immune system from mounting an overaggressive attack on them. Below, the role of immune cells in maintaining gut homeostasis is discussed further.

Microorganisms share a variety of cellular structural components that possess highly conserved features (e.g., similar or identical sequences in nucleic acids or proteins and conserved amino acid sequences in peptide antigens), a characteristic that is responsible for their designation as *pathogen-associated molecular patterns* or PAMPs. These PAMPs are essentially patterns in different microbes that produce the same immune-system-recognizable molecules. But as pointed out by Ausubel (2005), use of the term "pathogen" is inaccurate because these molecular patterns are shared by both commensals and pathogens; he proposed the use of the more apt term **microbe-associated molecular patterns** or MAMPs, an appellation that, surprisingly, has not been more widely adopted. Some common examples of MAMPs are the peptidoglycan polymer in cell walls of all bacteria, lipopolysaccharides (LPS) of Gram-negative bacteria, and lipoteichoic acids of Gram-positive bacteria. As noted in Chapter 2, enterocytes and immune cells possess a collection of so-called **pattern recognition receptors** (PRRs) that recognize, bind, and respond to the conserved patterns on MAMPs. Major examples of such PRRs are the surface-exposed toll-like receptors and the internalized NOD-like receptors. The activation of PRRs results in the induction of activities that protect epithelial surfaces and prevent microbial translocation across the epithelial barrier; key examples of such activities are the reinforcement of the epithelial tight junction and the secretion of antimicrobial peptides and mucins (Thaiss et al., 2016). Thus, the major force controlling the function of the gut immune system consists of sensory receptors on key immune cells that detect MAMPs.

Microbial metabolites such as short-chain fatty acids (SCFAs), secondary bile acids, and amino acid derivatives like indole are involved in regulating immune functions in the gut. The best characterized in this regard are the SCFAs acetate, propionate, and butyrate, the majority of which are end products of colonic fermentation of dietary fiber (see Chapter 2). SCFAs regulate the gut immune system through their specific interaction with a family of signal transduction proteins known as G-protein-coupled receptors (GPRs) such as GPR43 and GPR109a that occur on enterocytes and immune cells (Brestoff and Artis, 2013; Thaiss et al., 2016). SCFAs may have multiple effects on immune regulation, but in general, they are antiinflammatory and important for maintaining intestinal homeostasis (Kimura et al., 2014; Corrêa-Oliveira et al., 2016).

Immune cells play a critical role in determining the state of gut health. For example, a subpopulation of T cells called **regulatory T** (Treg) **cells** are important for maintaining gut homeostasis through their (antiinflammatory) ability to suppress the immune response to **antigens** (i.e., large molecules that can induce antibody production) derived from dietary components and the commensal microbiota. Gut SCFAs promote the differentiation of naive T cells to Treg cells (Furusawa et al., 2013; Arpaia et al., 2013; Smith et al., 2013). Furthermore, polysaccharide A, a capsule produced by the common gut commensal *Bacteroides fragilis,* also promotes Treg cell differentiation; notably, polysaccharide A was observed to prevent inflammatory conditions associated with colitis (Round and Mazmanian, 2010) and encephalomyelitis (Ochoa-Reparaz et al., 2010) in mice.

The relatively recently described **innate immune lymphocyte group 3 cell** (ILC3) is another important mediator of gut homeostasis. ILC3s produce the cytokine interleukin-22 (IL22) in response to the activation of their aryl hydrocarbon receptors by the metabolites produced when the microbiota metabolize tryptophan, such as indole-3-acetic acid (Lamas et al., 2016). IL22 receptors are expressed in numerous tissues in the body, including the mucosal epithelia, hepatocytes, and pancreatic cells (Sabat et al., 2014). In these tissues, IL22 performs a wide range of activities; in mice, depletion of ILC3s results in the systemic dissemination of gut commensal microbiota, a clear demonstration that IL22-producing ILC3s are essential for maintaining intestinal epithelial barrier function (Sonnenberg et al., 2012). ILC3s can also mediate immune surveillance to continuously maintain a normal microbiota. In mice, they were shown to facilitate early resistance against the pathogen *Citrobacter rodentium* through regulation of IL22 (Guo et al., 2015).

The above explanation shows gut homeostasis depends on a codependent relationship between the commensal microbiota and the intestinal immune system. As discussed below, a breakdown in this association may contribute to the development of localized inflammatory diseases like Crohn's disease or to conditions of systemic inflammation such as cardiovascular disease.

SIGNIFICANT MICROBIAL ENTERIC PATHOGENS

The best understood gut microbes are pathogens that are remarkably well adapted for gaining access to the gut environment by avoiding immune surveillance. The following are examples of some of the most important ones.

Foodborne Pathogens

The Centers for Disease Control and Prevention estimates that one in six Americans contracts a food-borne illness and 3000 die annually (Scallan et al., 2011). The World Health Organization, calling these diarrheal diseases, listed them as the eighth leading cause of death worldwide in 2015, responsible for nearly 1.4 million deaths, mostly in low- and middle-income countries (WHO, 2017). *Salmonella enterica*, *Campylobacter* species, *Escherichia coli* O157:H7, and *Listeria monocytogenes* are among the most frequent causes of bacterial gut infections. They share several characteristics in common. All appear to be highly specialized gut pathogens that almost exclusively utilize the oral cavity as a portal of entry—usually, through contaminated food. Curiously, all were first described as gut pathogens fairly recently (i.e., in the 1970s and 1980s). All but *Listeria* are typically carried asymptomatically in animal reservoirs that include common domestic animals. *Listeria* is environmentally ubiquitous and may be isolated, for example, from soil and water; although it was first described in 1924, it was not recognized as a food-borne pathogen until 1981 (Cartwright et al., 2013).

The epidemiological work of Cohen and Tauxe provides a likely explanation for the increased incidence of food-borne infections that started in the 1970s and 1980s (Cohen and Tauxe, 1986). *Salmonella* species are associated with two main diseases, typhoid fever caused by *S. typhi* and nontyphoid salmonellosis caused by many different serotypes of *S. enterica*. Animal reservoirs of *S. typhi* are unknown, and typhoid fever (classically a waterborne disease) ceased to be a public health problem by the mid-twentieth century in industrialized countries, with the introduction of sanitary practices like water purification and sewage treatment. Nontyphoid salmonellosis, on the other hand, has supplanted it as a priority health problem. As noted above, unlike *S. typhi*, *S. enterica* is a common gut commensal in a variety of wild and domestic animals, including cattle, swine, and poultry. Cohen and Tauxe used molecular techniques to demonstrate these animal reservoirs were the source of *S. enterica* responsible for human infections through contaminated animal products like milk and beef. They proposed that the observed increased incidence of nontyphoid salmonellosis was related to changes in livestock husbandry associated with the transition to industrialized agriculture in the latter half of the 20th century. As gut commensals in animal reservoirs, *S. enterica* are shed in feces, and it is not difficult to imagine how these microbes spread throughout entire herds and flocks maintained in crowded conditions. The industrialization of food processing in order to generate a higher volume of product resulted in the inadvertent contamination of processed goods. Most infections caused by *Salmonella, Campylobacter*, and *E. coli* have been associated with products derived from their animal hosts, whereas *Listeria* (being widespread environmentally) has been associated with a wider range of products. A disturbing trend in recent years is the increasing occurrence of *Salmonella* and *E. coli* in nonanimal products, possibly indicating the environmental dispersal of these pathogens. High-volume agriculture is here to stay, so the food industry must aim to solve the contamination problem.

Cohen and Tauxe (1986) also noted the similarity between the drug-resistant patterns exhibited by *Salmonella* strains isolated from animal sources compared with human sources and that these patterns correlated directly with the spectrum of drugs to which the animals had been exposed. Animal feeds have been supplemented with subtherapeutic levels of antibiotics since the 1950s, when it was noted that they stimulated animal growth, and this practice endures to this day in North America despite the knowledge that this results in the selection of drug-resistant microbes (Davies and Davies, 2010). Bacteria, mostly nonpathogenic commensals, carrying a diverse collection of genes encoding antibiotic resistance, are ubiquitous in the mammalian gut. An important feature of many of these genes is that they can undergo **horizontal transfer**—that is, they can be passed through one of several mechanisms to completely different species, unlike vertical gene transfer involving passage from parent cell to offspring. Horizontal gene transfer is a common—possibly universal—phenomenon, and as a crowded convention of hundreds of bacterial species, the gut is a perfect place for it to occur. The types of genes that are horizontally transferred are not restricted to drug resistance and, significantly, may encode virulence mechanisms on genetic units that are embedded in a microbe's genome. In fact, all of the enteric pathogens discussed in this chapter have had their genomes altered by the incorporation of such genetic units that encode significant virulence factors (Zhang et al., 2016; Chung et al., 2016; Nieto et al., 2015). The evolution of a harmless commensal *E. coli* into a deadly pathogen like strain O157:H7 involved the horizontal acquisition of numerous genes encoding an array of virulence factors, including its hallmark Shiga toxin (Sadiq et al., 2014).

These enteric pathogens all exhibit the ability to cross the gut epithelial barrier. *Salmonella* and *Listeria* each have their own unique ways to invade and replicate within epithelial cells (Malik-Kale et al., 2012; Pizarro-cerda and Ku, 2012). Enterohemorrhagic *E. coli* (EHEC) attaches to the follicle-associated epithelium of the Peyer's patch, where it exploits the unique translocation system of M cells to cross the epithelial barrier (Etienne-Mesmin et al., 2011). The translocated *E. coli* are then phagocytosed by macrophages, in which they replicate, produce Shiga toxin, and are released into the lamina propria upon death of the macrophage. Shiga toxin destroys endothelial cells of blood vessels in the gut, kidneys, and lungs, causing hemorrhaging—the basis for the term enterohemorrhagic in EHEC. *Campylobacter* uses its high degree of motility and corkscrew cellular morphology to literally bore its way into the gut mucosa to initiate infection (Bolton, 2015). *Campylobacter jejuni* also penetrates the gut epithelial barrier, but the precise mechanism is unclear (Backert et al., 2013). A recent report indicates that cellular invasiveness is enhanced by a subset of proteins secreted by *C. jejuni* (Scanlan et al., 2017).

Clostridium Difficile Infection

Clostridium difficile is the most frequent cause of hospital-acquired infections in the United States (Lessa et al., 2015). A key feature of *C. difficile* is its ability to differentiate into a dormant, nonmetabolizing form known as an **endospore**; in the laboratory context, this process occurs when conditions are not conducive for growth (Abt et al., 2016). Endospores are remarkably resistant to toxic chemicals, desiccation, and harsh physical treatments like heat and radiation and are designed to persist for extended periods (Gil et al., 2017). They represent a means of survival and environmental dispersal; they undoubtedly complicate disinfection procedures in hospitals. When conditions become favorable, endospores germinate to yield normal metabolizing cells. For *C. difficile*, two conditions are required for endospore germination. The environment must be oxygen-free (because *C. difficile* is an anaerobe), and there must be a source of certain primary bile acids. The digestive tract satisfies both requirements. *C. difficile* endospores possess specific receptors that are activated by bile acids to initiate germination.

A health-associated microbiota includes species that inhibit the growth of *C. difficile* (Fig. 4.1). *Clostridium difficile* infection (CDI) is frequently the result of factors like treatment with antibiotics or proton pump inhibitors that kill these beneficial species. *C. difficile* produces two major toxins called TcdA and TcdB (Abt et al., 2016). These toxins destroy the epithelial tight junctions, causing increased intestinal permeability and inflammation, resulting in symptoms ranging from watery diarrhea to pseudomembranous colitis. In animal models, TcdB proved to be the major virulence factor (Carter et al., 2015).

The usual CDI treatment, a combination of metronidazole and vancomycin, is designed to kill *C. difficile*, but recurrence of CDI after this treatment poses a challenge (Vindigni and Surawicz, 2015). Fecal microbiota transplantation (FMT) has gained interest as a treatment for recurrent CDI, and its reported efficacy rates in clinical trials have been as high as 90% (van Nood et al., 2013; Kelly et al., 2016). The object of FMT is to restore the microbiota to its original health-associated state, ostensibly because the gut microbial community has been severely depleted or disturbed. Microbial therapies involving defined probiotic mixtures are a more systematic approach for achieving the same objective; in this case, scientists would assemble a defined mixture of species isolated from feces that is experimentally effective in restoring the gut microbiota (Almeida et al., 2016). Using this method, a preparation containing 33 gut microbial species was shown to cure two cases of hypervirulent CDI by restoring the normal microbiota

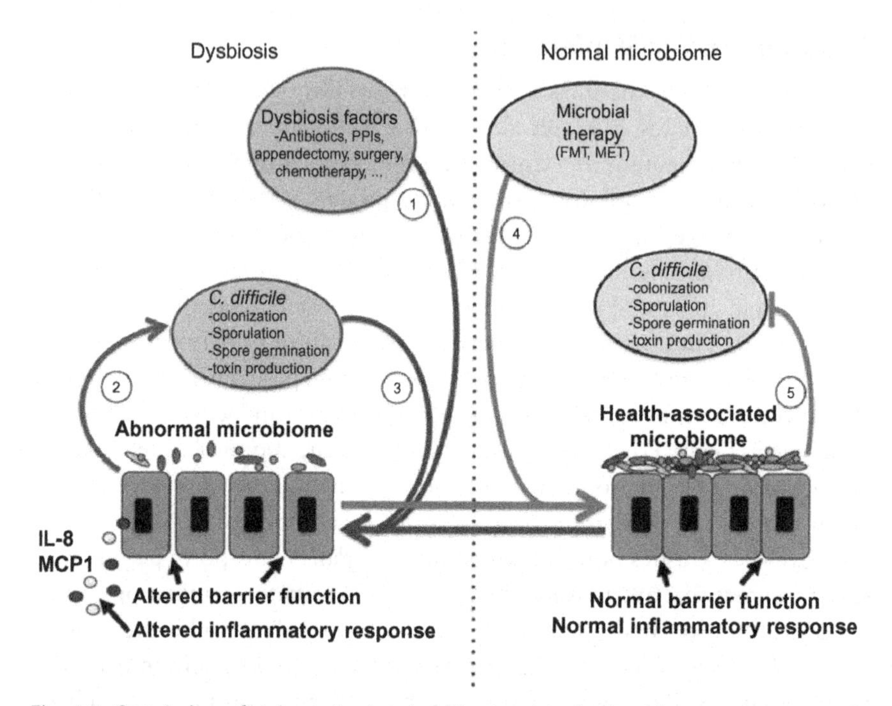

Fig. 4.1 Certain beneficial species in a health-associated microbiota produce secondary bile acids that inhibit the growth of *C. difficile* (right). A number of factors, such as treatments with antibiotics and proton pump inhibitors, kill the beneficial species. This results in production of toxins that increase epithelial barrier permeability and induce inflammation, leading to *C. difficile* infection (left). "Microbial therapy" treatments are promising for recurrent CDI. *(From Almeida, R., Gerbaba, T., Petrof, E. O., 2016. Recurrent Clostridium difficile infection and the microbiome. J. Gastroenterol. 51(1), 1–10. Copyright Springer Japan 2015, with permission of Springer.)*

within a few days (Petrof et al., 2013). In designing such therapies, an important consideration is the identification of secondary bile acids (SBAs) as the inhibitors of *C. difficile* growth in the gut (Theriot et al., 2015). Primary bile acids produced by the liver and secreted into the small intestine are modified to SBAs by certain members of the microbiota that are killed by antibiotic treatment.

Norovirus

Norovirus is the leading overall cause of gastroenteritis globally, responsible for an estimated 125 million cases per year (Kirk et al., 2015). Noroviruses have been classified into seven genogroups, three of which are known to cause human infections (de Graaf et al., 2017). These viruses are highly contagious, partly because of their environmental stability, and often cause large outbreaks in group settings such as cruise ships, hospitals, and care homes. Virus transmission may be from person to person, through contaminated food or water, or from environmental sources. Identifying the source of the virus is critical in outbreaks but is complicated in the case of norovirus because multiple modes of transmission may be involved (Verhoef et al., 2015). New laboratory culture methods and mouse models have significantly increased understandings of norovirus infection (Baldridge et al., 2016). In isolating gut viruses, fecal samples are usually filtered to remove bacteria prior to attempting virus culture. It was discovered that norovirus could not be cultured from filtered fecal samples but could be cultured if the filtration step was omitted (Jones et al., 2014). The presence of certain gut bacteria, *Enterobacter cloacae* being one, was required for successful isolation of norovirus. *Enterobacter cloacae* fortuitously possesses a human blood group antigen in its cell wall that also occurs on human erythrocytes and gut epithelial cells; norovirus binds to this antigen on *E. cloacae*, and this is hypothesized to stimulate the attachment of the virus to its host cells. Gut epithelial cells and immune cells such as macrophages, dendritic cells, and especially B cells have now been identified as hosts that support norovirus replication. The availability of viral culture methods will undoubtedly lead to a better understanding of this important gut pathogen.

COMPLEX DISEASES LINKED TO THE GUT MICROBIOTA

Asthma and Allergy

The most common forms of allergies occur when a foreign antigen, in this case referred to as an **allergen**, activates a specific subset of T cells called Th2 cells. The allergen-activated Th2 cell in turn induces plasma cells to produce allergen-specific antibodies belonging to a particular class known as **immunoglobulin E** (IgE). The allergen-specific IgEs attach to receptors on mast cells, and when this happens, the mast cells are said to be **sensitized**. Mast cells are tissue-dwelling immune cells that are characterized by numerous intracellular granules filled with pro-inflammatory chemicals, one of which is histamine. Allergens that are subsequently introduced attach to the cell-bound IgEs, and this causes the release of the chemicals stored in the mast cell granules. This is followed by an inflammatory response to the released chemicals.

As noted above, the gut immune system tolerates the presence of an immense number of commensal microbes in the digestive tract. Individuals ingest a large number of foreign materials through the daily diet, and the gut immune system, likewise, usually tolerates these materials—a phenomenon called **oral tolerance**. On the other hand, food allergies are common examples of IgE-mediated allergic responses, and these occurrences (involving a loss of oral tolerance) are directly related to dysfunction of the gut homeostasis mechanism (Adami and Bracken, 2016). The relationship between oral tolerance and the gut microbiota is clearly established by the observation that germ-free mice do not develop oral tolerance; more details on gut microbiota and oral tolerance will emerge with further study.

Asthma is an IgE-mediated airway allergy to a substance that gains entry through the respiratory tract (Adami and Bracken, 2016). Fig. 4.2A compares a nonasthmatic airway with an asthmatic airway. The major distinction is the decreased opening of the asthmatic airway due to enlargement of smooth muscle. In the healthy airway, alveolar macrophages patrol the airway lumen, and many immune cell types occupy the lung tissue—the most

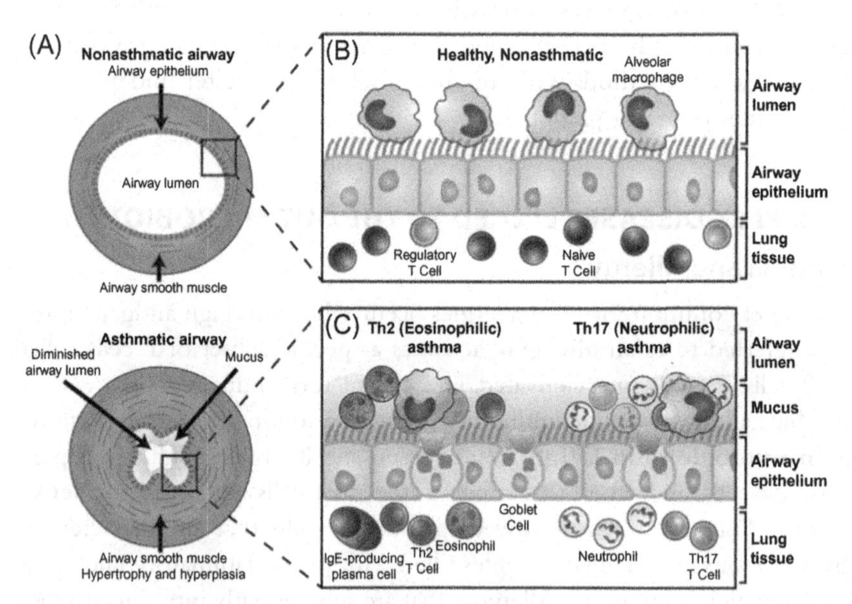

Fig. 4.2 Comparison of a nonasthmatic and an asthmatic airway. (A) Cross-sectional view. (B) Features of a healthy airway. (C) Features of two forms of asthmatic airways: Th2 or eosinophilic asthma (left) and Th17 or neutrophilic asthma (right). *(From Adami, A. J., Bracken, S. J., 2016. Breathing better through bugs: asthma and the microbiome. Yale J. Biol. Med. 89(3), 309–324.)*

significant being naive (nonactivated) T cells and Treg cells (Fig. 4.2B). As noted above, Treg cells suppress the activities of immune cells. There are two forms of asthma. The most common is Th2 asthma (Fig. 4.2C, left), characterized by the presence of allergen-specific Th2 cells that program plasma cells to produce allergen-specific IgE; they also secrete cytokines that attract inflammatory eosinophils to the site. The second form of asthma is not mediated by Th2 cells and IgE; instead, it is driven by another subset of T cell known as Th17, noted for its production of the pro-inflammatory cytokine IL17 (Fig. 4.2C, right). This form of asthma is known as Th17 or neutrophilic asthma; the latter term is used because neutrophils are recruited to the respiratory tissue.

The incidence of allergies has increased significantly over the past four to five decades, especially in high-income countries. In an effort to explain this trend, Strachan proposed the **hygiene hypothesis**, which linked the increasing incidences of certain diseases like allergies to "higher standards of personal cleanliness" that effectively decreased exposure to infectious microbes (Strachan, 2000). Evidence supporting this hypothesis (or the slightly modified "old friends" hypothesis from Rook and Brunet, 2005) continues to accumulate; for example, see the described asthma incidence in Amish and Hutterite populations in Chapter 5. Asthma and allergies often develop in early life and are associated with gut microbiota dysbiosis. A recent study of 319 healthy infants found the development of asthma was most likely to occur in infants who exhibited gut microbiota differences in the initial 100 days of life (Arrieta et al., 2015). Moreover, those at risk showed a significantly decreased abundance of four bacterial genera, *Lachnospira*, *Faecalibacterium*, *Rothia*, and *Veillonella*. A causal relationship between these bacteria and asthma development was demonstrated by inoculation of germ-free mice with these microbes.

Inflammatory Bowel Disease

Research supports a definite role for the gut microbiota in the pathogenesis of inflammatory bowel disease (IBD), including both Crohn's disease (CD) and ulcerative colitis (UC). According to two leading researchers in the field, Sartor and Wu, CD and UC currently "appear to result from overly aggressive T-cell-mediated immune responses to specific components of the intestinal microbiota in genetically susceptible hosts, with disease initiated and reactivated by environmental triggers" (Sartor and Wu, 2017). Fig. 4.3 shows how genetic and environmental factors that affect the intestinal microbiota may play a role in IBD.

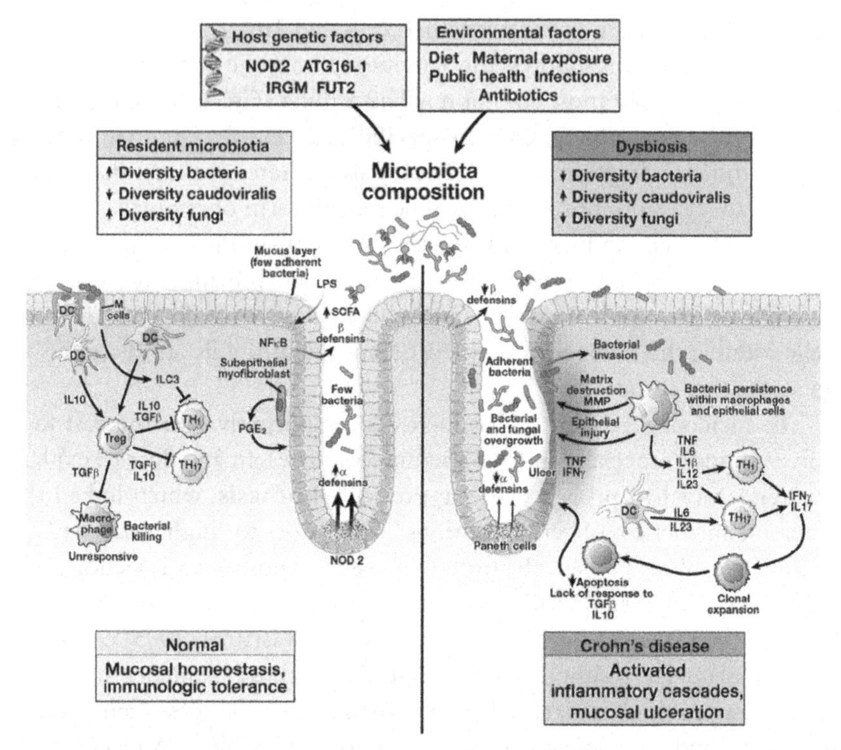

Fig. 4.3 Genetic and environmental factors potentially affecting microbiota composition, which plays a role in IBD pathogenesis. *(From Sartor, R. B., Wu, G. D., 2017. Roles for intestinal bacteria, viruses, and fungi in pathogenesis of inflammatory bowel diseases and therapeutic approaches. Gastroenterology, 152(2), 327–339. Copyright 2017, with permission from Elsevier.)*

Gut mucosal microbial diversity is reduced in IBD (Walker et al., 2011; Ott et al., 2004), most dramatically in CD. Differences in species composition compared with healthy individuals are also repeatedly found; in both CD and UC, a general trend is observed toward the depletion of SCFA producers like *Eubacterium*, *Roseburia*, and especially *Faecalibacterium prausnitzii*—an anaerobic, antiinflammatory butyrate producer. An alteration of fungal microbiota in IBD was also recently reported: an increased Basidiomycota/Ascomycota ratio compared with healthy individuals and a decreased proportion of *Saccharomyces cerevisiae* as well as an increased proportion of *Candida albicans* (Sokol et al., 2016).

Consistent microbial signatures of either CD or UC are largely elusive, but there may be biomarker potential when it comes to CD. Multiple studies reveal that members of Enterobacteriaceae, specifically *E. coli* adherent-invasive strains, are increased in the intestines of those with CD, while *F. prausnitzii* is decreased. In addition to lower levels of SCFAs, functional changes are also present: major changes in oxidative stress pathways and decreases in carbohydrate metabolism and amino acid biosynthesis (Wright et al., 2015). A recent study that analyzed the microbiota composition of over 2000 patients with and without IBD from four countries (Spain, Belgium, the United Kingdom, and Germany) showed distinct microbiome markers of CD, independent of geographic region (Pascal et al., 2017): researchers saw the loss of several butyrate-producing microorganisms such as *Faecalibacterium*, Christensenellaceae, *Methanobrevibacter*, and *Oscillospira*. An algorithm based on eight genera allowed these researchers to identify individuals with CD.

A mechanistic study from France showed how host genes may affect both the composition and function of the gut microbiota, which in turn influence the production of metabolites and IBD-like inflammation. Mice without the caspase recruitment domain family member 9 (CARD9) gene—which were susceptible to colitis—showed a different microbiota composition than mice with the gene, and when genetically normal germ-free mice were colonized with microbiota from CARD9-deficient mice, they were more susceptible to gut inflammation. This was associated with decreased levels of IL22 in the colon; so compositional and functional gut microbiota alterations resulting from CARD9 deficiency were responsible for defective IL22 production from gut immune cells, leading to inflammation (Lamas et al., 2016). This provides a new perspective on how genes could influence disease susceptibility through the microbiota, and since the CARD9 gene has also been implicated in human IBD (Rivas et al., 2011), this work may provide a direction for future therapeutics.

Metabolic Syndrome and Obesity

In the past decade, researchers have greatly advanced knowledge about the connection between the gut microbiota and metabolic disease, including overweight/obesity. Metabolic syndrome encompasses a complex of symptoms that include loss of glycemic control as well as dyslipidemia, hypertension, and adiposity; obesity involves an excess of adipose tissue and is closely related to metabolic syndrome. Various links between these conditions and the human gut microbiota are being explored.

Human metabolic diseases are often associated with decreased diversity and functional richness of the gut microbiota (Wu et al., 2015). When it comes to obesity, a 2004 study (Bäckhed et al., 2004) demonstrated that slim germ-free mice, when transplanted with a normal microbiota, increased their body fat by 60% and increased their insulin resistance while reducing food intake; researchers proposed (Turnbaugh et al., 2006) the microbiota of some individuals may be more efficient than that of others at extracting energy from a given diet. Then, a 2013 humanized mouse study (Ridaura et al., 2013) showed human fecal microbiota could transfer obesity to germ-free mice; when researchers took fecal microbiota from female genetically identical twin pairs consisting of one lean member and one obese member, the mice receiving the gut microbiota of the obese individual gained more weight than the mice receiving the microbiota from the lean member. Cohousing the mice, however, kept both groups of mice lean (although this was dependent on diet).

Translating these dramatic results from mice into humans has not been straightforward. Initial efforts (Turnbaugh et al., 2009) seemed to converge on the idea that the gut microbiota of lean and obese individuals differed at the phylum level, with an increased ratio of Firmicutes to Bacteroidetes in obese individuals. But gut microbial signatures of obesity proved inconsistent from study to study. Recent analyses (Walters et al., 2014; Sze and Schloss, 2016) showed obesity is only weakly associated with particular bacterial groups, with studies failing to support even the relevance of the ratio of Firmicutes to Bacteroidetes in obesity. Recent work has suggested complex consortia of bacteria in humans could play a role in metabolic perturbations: 22 bacterial species and four operational taxonomic units (OTUs) were either positively or negatively correlated with metabolic syndrome traits in a study of 310 individuals with varying body mass indexes (BMIs) (Zupancic et al., 2012).

Some scientists have focused on key species of relevance in affecting metabolic parameters—in particular, *Akkermansia muciniphila*, a mucin-degrading bacterium that resides in the nutrient-rich mucus layer of the gut. Previous work showed that *A. muciniphila* levels decreased in obese and type 2 diabetic mice, and treatment with these bacteria reversed high-fat diet-induced metabolic disorders, including fat mass gain, metabolic endotoxemia, adipose tissue inflammation, and insulin resistance (Everard et al., 2013). *A. muciniphila* was able to control mucus production by the host and restore mucus layer thickness in mice with high-fat diet-induced obesity, thereby reducing gut permeability. This led to the hypothesis that *A. muciniphila* engages in cross talk with the intestinal epithelium to control inflammation and gut barrier function in the pathophysiology of obesity. And while the role of *A. muciniphila* is less certain in humans, it is depleted in those with several metabolic and inflammatory disorders; one study found that in subjects undergoing a calorie restriction treatment for obesity, those with higher levels of these bacteria exhibited the best metabolic status and clinical outcomes (Dao et al., 2016).

A recent study showed a pasteurized (heat-killed) version of *A. muciniphila* reduced fat mass development and insulin resistance in mice and modulated intestinal energy absorption and the host urinary metabolome (Plovier et al., 2016). Researchers attributed these effects to a protein called Amuc_1100*, found on the outer membrane of the bacterium, which appeared to interact with toll-like receptor 2. Others have shown this protein led to high levels of IL-10 and improvements in gut barrier function (Ottman et al., 2017). The antiinflammatory activities of this bacterial protein presumably explain its effects.

One leading theory on the pathogenesis of obesity emphasizes a close link between the metabolic and immune systems via the gut microbiota. A body of work shows that when the gut microbiota absorb bacterial **lipopolysaccharide** (LPS), an outer membrane component of Gram-negative bacteria, the resultant increase in intestinal permeability can lead to the release of bacterial endotoxin (i.e., LPS) through the damaged gut, resulting in metabolic endotoxemia. Activation of pro-inflammatory cytokines is observed, leading to the chronic low-grade inflammation known to be implicated in obesity (Khan et al., 2016). This process is shown in Fig. 4.4.

The known connection between obesity and gut microbiota reinforces the notion of obesity as a disorder of complicated etiology, countering the stigmatizing notion that it is attributable to poor lifestyle choices. A recent endocrinology position paper argued for a reframing of obesity by naming it "adiposity-based chronic disease" (ABCD) (Mechanick et al., 2016).

Type 2 Diabetes

Related to the links described above, gut microbes are emerging as key players in the development of insulin resistance. A 2012 large-scale study found the gut microbiota of Chinese individuals with type 2 diabetes differed from that of controls; those with type 2 diabetes showed a decrease in the abundance of some butyrate-producing bacteria and an increase in opportunistic pathogens. Further, they had an enrichment in microbial genes for sulfate reduction and oxidative stress resistance (Qin et al., 2012). Numerous other studies have found compositional microbiota differences between those with type 2 diabetes and healthy controls, and while there is no single compositional or functional feature that signals this form of

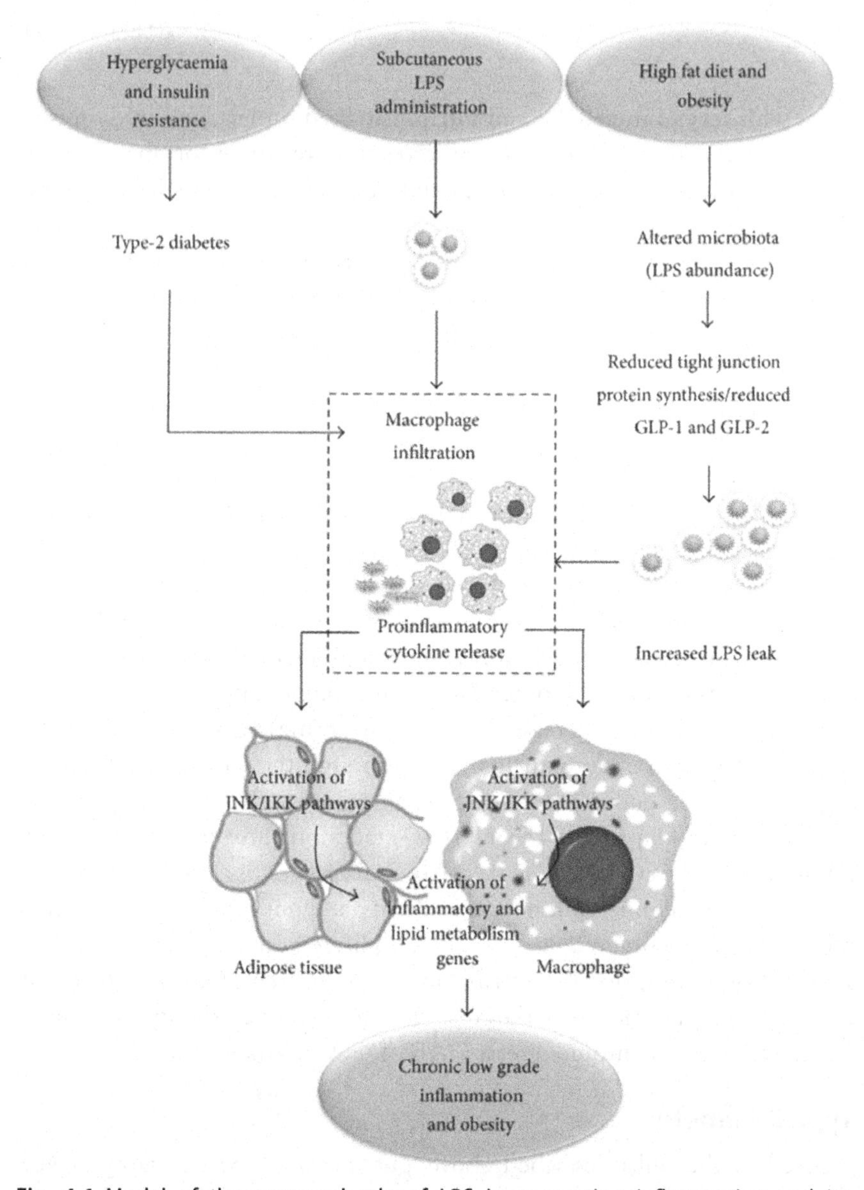

Fig. 4.4 Model of the proposed role of LPS in generating inflammation and its relationship with obesity. Altered mucosal barrier function due to reduced expression of glucagon-like peptides 1 and 2 (GLP-1 and GLP-2) leads to altered mucosal function and reduced synthesis of tight junction proteins, zonula occludens-1 and zonula occludens-2 (ZO-1 and ZO-2), increasing gut permeability. This allows LPS to enter systemic circulation, leading to the release of pro-inflammatory cytokines. These in turn result in the activation of a family of kinases JNK and IKK (inhibitor of NFkB kinase) that increase the expression of inflammatory and lipid metabolism genes. Subcutaneous administration of LPS, hyperglycemia, and insulin resistance induces the same pathway by increasing endoplasmic reticulum and mitochondrial stress. Type 2 diabetes, hyperglycemia, and insulin resistance also cause macrophage infiltration and inflammatory cytokine release, with the same effects as a high-fat diet (HF). *(From Khan, M. J. et al., 2016. Role of gut microbiota in the aetiology of obesity: proposed mechanisms and review of the literature. J. Obes. 2016, 7353642.)*

diabetes, the bacteria that differ between groups are those influencing inflammation and energy homeostasis (Caricilli and Saad, 2013).

Low-grade systemic inflammation might drive the metabolic changes that lead to both insulin resistance and obesity. What is unknown to date is the factor(s) that initially trigger the inflammation. The bacteria themselves could play a role: if SCFA producers are reduced, the intestinal barrier may be impaired in a way that facilitates bacterial translocation. As described above, this may lead to increased plasma LPS and stimulation of an inflammatory response, cytokine production, and chemokine-mediated recruitment of acute inflammatory cells, resulting in metabolic endotoxemia (Cani et al., 2007). Bile acids are also under investigation in the pathogenesis of type 2 diabetes, since in the digestion of dietary lipids they may act as signaling molecules in the context of energy, glucose, and lipid metabolism (Prawitt et al., 2011).

Cardiovascular Disease

Cardiovascular disease (CVD) refers to conditions involving narrowed or blocked blood vessels that can lead to heart attack, angina, or stroke. Changes in gut microbiota structure and function have been observed in those with symptomatic atherosclerosis: one study found an enrichment in the genus *Collinsella* and a depletion of *Roseburia* and *Eubacterium* compared with healthy individuals, with gut metagenomes enriched in genes encoding peptidoglycan synthesis (Karlsson et al., 2012). Transplantation of gut microbiota from human donors with hypertension into germ-free mice was recently shown to elevate blood pressure in the recipient mice, showing a causal role of the microbiota (Li et al., 2017).

A study of nearly 900 volunteers showed a connection between gut microbiota and risk factors for human CVD. The study found 34 bacterial taxa associated with BMI and blood lipids, which are two important CVD risk factors. Microbiota explained 4.5% of variance in BMI, 6% in triglycerides, and 4% in high-density lipoproteins (HDL) but did not appear relevant to low-density lipoproteins (LDL). By accounting for 4.5% of BMI variance, microbiota data could be a more powerful predictive tool than human genetic data, which explain 2.1% of the variance. Plugging this information into a new model of CVD risk, authors could explain up to 25.9% of HDL variance. They concluded the gut microbiota plays a role in variation of BMI and blood lipids, supporting the idea that it could be targeted for the management of metabolic syndrome (Fu et al., 2015).

Work from Stanley Hazen's lab has brought forward a clear link between diet, gut microbes, and health by helping identify the proatherosclerotic metabolite **trimethylamine-N-oxide** (TMAO) as a risk factor for CVD. The link between red meat consumption and CVD risk is well-known. Lecithin, choline, betaine, and L-carnitine are trimethylamine (TMA)-containing dietary compounds that are particularly abundant in red meat. Upon ingestion, host gut microbes take in TMA-containing compounds and release TMA, which is then metabolized by enzymes in the liver to form TMAO (Liu et al., 2015). Gut microbes are essential for TMAO generation (Koeth et al., 2013), and it is the increase in TMAO, rather than the increase in levels of TMA-containing compounds in the gut, that serves as an independent risk factor for CVD (Wang et al., 2014).

Focusing in on L-carnitine, the researchers found that antecedent dietary habits (i.e., red meat consumption) might influence individuals' capacity to generate TMAO from L-carnitine; omnivores produced more of the compound than vegans or vegetarians when they ingested sources of L-carnitine. Furthermore, plasma L-carnitine levels predicted an increased risk for CVD and major adverse cardiac events (myocardial infarction, stroke, or death)—but this was only the case for subjects with high TMAO levels (Koeth et al., 2013).

In another study, researchers found that after a phosphatidylcholine challenge, healthy participants showed time-dependent increases in levels of TMAO. But after antibiotics, plasma TMAO levels were temporarily suppressed. In a follow-up with over 4000 patients undergoing elective coronary angiography, plasma TMAO levels predicted major adverse CV events even after adjusting for traditional risk factors (Tang et al., 2013).

Type 1 Diabetes

Type 1 diabetes is a chronic autoimmune disease that occurs when insulin-producing beta cells in the pancreas are destroyed in genetically susceptible individuals. In humans with preclinical type 1 diabetes, the dominant phylum in the gut microbiota is Bacteroidetes. These individuals also show fewer butyrate-producing bacteria, reduced bacterial and functional diversity, and low community stability (Knip and Siljander, 2016). An informative integrated multiomics study of 20 people from four families with multiple cases of type 1 diabetes found those with diabetes had differences in the relative abundances of various pancreatic enzymes in the stool, which correlated with the expression of microbial genes that included thiamine synthesis and glycolysis. While no consistent taxonomic changes associated with type 1 diabetes were observed, several microbial populations appeared to contribute to the functional differences (Heintz-Buschart et al., 2016).

Children with beta-cell autoimmunity show lower lactate-producing and butyrate-producing species in the gut, with less of two dominant *Bifidobacterium* species—*B. adolescentis* and *B. pseudocatenulatum*—and more bacteria in the *Bacteroides* genus (de Goffau et al., 2013). Predictive value may exist in the gut microbiota: in a study of 33 infants who were genetically predisposed to type 1 diabetes, those who progressed to disease showed a drop in alpha diversity between seroconversion and disease diagnosis, with an increase in bacteria, gene functions, and serum/stool metabolites that were linked to inflammation (Kostic et al., 2015). Because the gut microbiota changes occur after the appearance of autoantibodies, the gut microbiota may be involved in the progression from beta-cell autoimmunity to clinical disease, rather than in initiation of the disease.

Liver Disease

More than two-thirds of the blood directed to the liver comes from the gut via the portal venous system (Ianiro et al., 2016). As a result, the liver is exposed to a great number of bacterial components and metabolites (Tilg et al., 2016). Growing evidence is implicating the gut microbiota in the pathogenesis of several liver diseases.

Nonalcoholic fatty liver disease (NAFLD) includes a range of liver conditions related to excess fat stored in liver cells and is the most common cause of chronic liver disease worldwide. Pathologies range from nonalcoholic fatty liver (NAFL) to steatohepatitis (NASH), possibly progressing to fibrosis (Del Chierico et al., 2014).

Gut microbiota play a role in the regulation of hepatic lipogenesis (Delzenne and Kok, 1998), and indeed, changes in gut microbiota composition are observed in NAFLD. One study found *Lactobacillus* species and selected members of the phylum Firmicutes (Lachnospiraceae; genera included *Dorea*, *Robinsoniella*, and *Roseburia*) were increased in those with NAFLD and obesity. A spike in fecal ester volatile organic compounds was associated with these compositional shifts (Raman et al., 2013). Another line of work has found an association between the existence of NASH and decreased levels of Bacteroidetes (Mouzaki et al., 2013). Gut microbial clues about NAFLD severity have also been discovered: Boursier and colleagues found shifts in both gut microbiota structure and function that were associated with NAFLD severity (Boursier et al., 2016), with *Bacteroides* being associated with NASH and *Ruminococcus* with significant fibrosis.

Gut microbiota may also contribute to the development of hepatic encephalopathy (HE)—a deterioration in brain function (with possible altered consciousness or coma) resulting from liver failure. Gut microbes likely influence the development of HE by producing ammonia, which may elicit a systemic inflammatory response that has effects on the brain (Shawcross et al., 2010). A study in a mouse model of cirrhosis confirmed gut microbiota changes drive the neuroinflammatory and systemic inflammatory responses occurring as a result of the disease (Kang et al., 2016).

Alcoholic liver disease (ALD) covers a spectrum of disorders related to excessive alcohol consumption over time. As early as 1984 (Bode et al., 1984), it was observed that those with chronic alcohol abuse showed small intestinal bacterial overgrowth. Current understandings of disease pathogenesis postulate that alcohol-induced disturbance of gut permeability leads to systemic circulation of LPS (Szabo, 2015); gut microbiota alterations in alcoholics appear to correlate with high levels of serum LPS (Mutlu et al., 2012). The impaired intestinal epithelial barrier may occur before these gut microbiota alterations.

Cirrhosis, irreversible liver scarring, occurs as a consequence of many chronic liver diseases. Data from the European MetaHIT project showed a large number of gut bacterial genes differing in abundance between those with cirrhosis and those who were healthy. Most of these species were of buccal origin (Qin et al., 2014), showing that liver cirrhosis may be characterized by oral microbiota being overrepresented in the lower gastrointestinal tract. Other studies have shown gut microbiota differences in cirrhosis; at the phylum level, the proportion of Bacteroidetes was found to be reduced, while Proteobacteria and Fusobacteria were enriched. (Chen et al., 2011). A group of researchers used an index of gut microbiota dysbiosis in those with cirrhosis and found the degree of dysbiosis to be correlated with endotoxin and with the progressive severity of disease (Bajaj et al., 2014).

Necrotizing Enterocolitis

Necrotizing enterocolitis (NEC) occurs in some preterm infants when portions of the bowel undergo necrosis (tissue death); gut microbiota is being investigated as a factor in the pathogenesis of this serious disease. A meta-analysis showed that, before NEC onset, the fecal microbiome of infants showed increased relative abundances of Proteobacteria, with lower relative abundances of Firmicutes and Bacteroidetes, compared with unaffected infants (Pammi et al., 2017).

Irritable Bowel Syndrome

Irritable bowel syndrome (IBS) is a highly prevalent condition of chronic abdominal pain, bloating, flatulence, and/or alterations in bowel habits in the absence of an identifiable physical cause. Many studies, including some dating back more than 30 years (Balsari et al., 1982), have identified differences in gut microbiota composition between those with IBS and healthy controls, with alterations that include reduced diversity, different proportions of specific bacterial groups (e.g., decreased lactobacilli and bifidobacteria and a trade-off of aerobes for anaerobes), and higher temporal instability. Bacteria in the mucosa are also more abundant in those with IBS (Distrutti et al., 2016). It is also known that IBS can occur after an acute gastrointestinal infection (Thabane and Marshall, 2009), leading some to posit that a gut microbiota severely altered by infection is involved in the pathogenesis of some forms of IBS. Geographic patterns may vary for this condition, as shown by one analysis that found the gut microbiota signatures of those with and without IBS were different in people from China versus other regions of the world (Zhuang et al., 2017).

The gut microbiota may emerge as a long-awaited biomarker for the functional syndrome of IBS. A study of Swedish adults with IBS found no differences in fecal microbiota composition compared with healthy controls, but through a novel machine-learning approach, researchers identified a microbial signature for IBS severity that consisted of 90 bacterial operational taxonomic units (Tap et al., 2017). Validation of this signature in people from various geographic groups remains to be seen.

Cancer

Recently, links have been made with the gut microbiota and colorectal cancer (CRC). Heredity undoubtedly contributes to CRC, but many studies have been devoted to identifying relevant environmental influences. The gut microbiota, modified in particular by diet, is emerging as an important factor. Studies correlate changes in fecal microbiota composition in individuals with CRC compared with healthy controls; these generally show reductions in some butyrate-producing species. Moreover, fecal transplants from mice with CRC into germ-free mice lead to increased tumorigenesis (Zackular et al., 2013).

The capacity of diet to modulate gut microbiota may be important in the link between gut microbiota and CRC, as shown by a human study in which a high-fat, low-fiber diet consumed for two weeks led to an increase in mucosal biomarkers of CRC risk (O'Keefe et al., 2015). The mechanism linking diet with colon cancer may have to do with consistently high exposure of gastrointestinal tract cells to bile acids (occurring in those with a high dietary fat intake) (Ajouz et al., 2014).

Researchers are also exploring how the preexisting gastric microbiota interacts with the presence of *H. pylori* to potentially influence the risk of gastric diseases and cancer (Touati, 2010), since *H. pylori* is strongly associated with gastric cancer and is considered a human carcinogen (O'Connor et al., 2017).

Malnutrition

Children with severe acute malnutrition (SAM) may receive therapeutic food interventions, but these often incompletely restore healthy growth. An important study of Bangladeshi children with SAM showed that they had "immature" gut microbial communities compared with others of their chronological age, and the immaturity did not resolve when they received therapeutic foods (Subramanian et al., 2014). This led to the idea that the gut microbiota is a key factor in determining the growth of malnourished children. The immature microbiota from SAM-affected children conferred the impaired growth phenotype when transplanted into germ-free mice, showing microbiota differences are causally related to the effects of malnutrition (Blanton et al., 2016). Mechanistic work to determine growth-promoting bacteria is ongoing; intestinal strains of *Lactobacillus plantarum* were recently found to promote growth in mice with chronic undernutrition (Schwarzer et al., 2016).

Celiac Disease

Celiac disease (CD) is characterized by an immune reaction triggered by dietary gluten, which is found in wheat, barley, rye, and other dietary constituents. CD is linked in a preliminary way with gut microbiota alterations: one study reported that individuals with CD experiencing symptoms despite adherence to a gluten-free diet showed lower microbial richness in the duodenum, with a higher relative abundance of Proteobacteria and a lower abundance of Bacteroidetes and Firmicutes (Wacklin et al., 2014).

A mechanistic study recently showed how bacteria isolated from the small intestines of CD patients (*Pseudomonas aeruginosa*) participated in gluten metabolism, generating a distinct pattern of gluten peptides upon digestion that was associated with increased immunogenicity. Bacteria from those without CD influenced gluten digestion differently, with decreased immunogenicity of gluten metabolism products (Caminero et al., 2016).

Brain-Related Conditions

Individuals with major depressive disorder appear to have differences in fecal microbial composition: one study found increased levels of Bacteroidetes,

Proteobacteria, and Actinobacteria, along with decreased Firmicutes, compared with nondepressed individuals. Researchers found a negative correlation between *Faecalibacterium* and the severity of depressive symptoms (Jiang et al., 2015).

When it comes to Parkinson's disease (PD), an initial study found intestinal microbiome alterations—reduced abundance of Prevotellaceae compared with controls—with the relative abundance of Enterobacteriaceae being positively associated with the degree of postural instability and gait difficulty (Scheperjans et al., 2015). Several other studies have found alterations of the gut microbiota in PD, with no consistent signature but with butyrate-producing (antiinflammatory) bacteria being reduced. In one intriguing study of a mouse model that emulated some features of PD, gut microbiota were found to be essential for the appearance of motor deficits, microglia activation, and synucleinopathies. When transplanted into mice, gut microbiota from humans with PD (but not from healthy individuals) enhanced the physical impairments in the rodent model of PD (Sampson et al., 2016).

Autism spectrum disorder (ASD) is a neurodevelopmental disorder characterized by abnormal social interaction and communication along with repetitive behaviors; it often co-occurs with gastrointestinal dysfunction. Subsets of those with ASD show alterations in gut microbiota, and investigation of the role of gut microbiota in ASD is in its early stages. In a study of children with functional gastrointestinal disorders with or without ASD, researchers found mucosal microbial features unique to those with ASD (an increase in members of the genus *Clostridium* and marked decreases in *Dorea*, *Blautia*, and *Sutterella*), which correlated with cytokine and tryptophan homeostasis (Luna et al., 2017). A causal link between maternal diet, gut microbiota, and behavior has been shown in animal models: in mice, a maternal high-fat diet induced a shift in microbiota that had a negative impact on offspring social behavior. Both the microbiota dysbiosis and the social deficits were transferable to germ-free mice and were preventable with exposure to the feces of mice whose mothers consumed a normal diet or with exposure to a single commensal strain (Buffington et al., 2016). Also, a gut-brain-microbiota connection in ASD was further supported by another study of a mouse model showing features of ASD that also exhibited gut barrier defects and microbiota alterations. Oral treatment of these mice with *B. fragilis* corrected the observed gut permeability, altered microbial composition, and ameliorated the behavioral deficits (Hsiao et al., 2013).

Other Conditions

Multiple sclerosis (MS) is a central nervous system disease characterized by lesions in the brain and spinal cord and inflammatory demyelination. While its pathogenesis is not fully known, links with gut microbiota are suspected. A limited number of human studies show diversity of the gut microbiota does not differ between those with MS and controls; however, certain enrichments and depletions in taxonomic groups may be present that suggest a pro-inflammatory milieu (Tremlett et al., 2017). For example, one study found a notable depletion of species belonging to Clostridia clusters XIVa and IV (Miyake et al., 2015). Also, in one mouse model of MS, the gut microbiota was found to be essential for triggering the autoimmune processes leading to disease (Berer et al., 2011).

Decreased gut microbial diversity has been observed in those with rheumatoid arthritis (RA), a progressive disease of joint inflammation. One study ($n = 40$) of those with RA found an increase in rare gut bacterial taxa (Actinobacteria) compared with healthy individuals. A correlation between gut microbiota and metabolic signatures was also observed, leading some to believe the gut microbiota could one day have predictive value in RA (Chen et al., 2016).

Systemic lupus erythematosus (SLE) is an autoimmune disease in which the body experiences a breakdown in tolerance toward self-antigens. While gut microbiota differences have been reported in SLE, a recent study went further and found that immune responses against certain gut bacteria could be involved in the lymphocyte overactivation and the Treg-Th17 transdifferentiation that has been observed in SLE-affected individuals. The researchers found a possible role for the bacterial phylum Synergistetes in the generation of protective humoral immune responses in SLE (López et al., 2016).

Human immunodeficiency virus (HIV) infection leads to acquired immune deficiency syndrome (AIDS) when severe immunodeficiency occurs. A remarkable expansion of the enteric virome has been shown to occur in HIV-infected individuals with severe immunodeficiency, along with less diversity and richness in the gut bacterial microbiota. One of the most noticeable changes in these individuals was reportedly an increase in inflammation-associated Enterobacteriaceae (Monaco et al., 2016). In the future, modulating the commensal microbial community may advance as a therapeutic adjunct for improving outcomes of HIV infection (Williams et al., 2016).

Preliminary evidence suggests not only gut microbiota alterations but also increased microbial translocation in myalgic encephalomyelitis/chronic fatigue syndrome (ME/CFS), a poorly understood chronic disorder characterized by extreme fatigue and various other symptoms. Those with ME/CFS showed a decrease in gut microbiota diversity with an increase in putative pro-inflammatory species and a reduction in antiinflammatory species (Giloteaux et al., 2016). Possibly, the altered gut microbiota may play a role in the increased translocation and inflammation in these individuals.

Defining a "Healthy" or "Unhealthy" Gut Microbiome

Many studies in this chapter have compared the gut microbiota in health versus disease. Altered gut microbiota composition and/or function is linked to a growing number of conditions, spanning the range from metabolic and liver disorders to some brain-related dysfunctions. Individuals in a known disease state are often found to have a statistically significant difference in gut microbiota composition compared with healthy individuals, and this difference can be called a **dysbiosis**. Some scientists have defined dysbiosis as a disruption of the complex gut microbial community (Petersen et al., 2014), but the precise set of bacteria that constitute a disruption is impossible to define for any individual. The word dysbiosis may be useful as a shorthand for referring to a disease-associated gut microbiota composition, but it cannot stand on its own as a diagnostic term.

The features that characterize a "healthy" gut microbiome, however, have been surprisingly elusive. Several notions have been proposed, as described below:

Balance Between "Good" Species and "Bad" Species

The simple concept of "good" versus "bad" bacteria dates back to the time of Pasteur and Metchnikoff, when scientists were aware of a limited number of culturable bacteria that lived in the human gut and could ostensibly be added or subtracted to affect health. Modern awareness of the enormously complex gut ecosystem, with hundreds of bacterial species vying for their own survival under changing conditions, makes this concept obsolete. While it remains true that the presence of pathogens generally has negative consequences, the effects of potentially deleterious species that may be present in the gut (e.g., *Bacteroides* species such as *B. fragilis*, certain forms of *E. coli*, and *Enterococcus* spp.) are found to depend on context. Furthermore, by now, it is clear that characterizing a healthy microbiome as a set of specific microorganisms is impossible (Lloyd-Price et al., 2016); no core set of microbial taxa is present in all healthy individuals.

Olesen and Alm recently argued that the common idea of dysbiosis as a microbiota "imbalance" will not contribute to clinically relevant insights (Olesen and Alm, 2016). Indeed, the concept of balance is difficult to define scientifically, and if a dysbiosis truly predicts a disease state, it will be clearly defined and will become an accepted biomarker.

Species Richness or Diversity

A common observation in the literature is increased species diversity and/or richness in the gut microbiota of healthy individuals. Counterexamples exist, however. As one example, recent work linked high gut microbial diversity with a longer colonic transit time and systemic circulation of potentially harmful protein degradation products (Roager et al., 2016). It could be that low diversity indicates poorer health, while high diversity does not always guarantee health. Thus, information about diversity alone is not sufficient to assess the health of the microbiota—or indeed of the host.

Functional Diversity

The number of microbial genes in the gut is positively associated with metabolic health (Le Chatelier et al., 2013) and perhaps other aspects of health as well. High diversity of metabolic and other molecular functions capable of being performed, regardless of the particular microorganisms present, is a promising marker of a healthy gut microbiota. From an ecological perspective, functional diversity may be a key factor in allowing an ecosystem to continue doing all the jobs it must do (Laureto et al., 2015). More research exploring functional diversity in the gut microbiota is required.

Stability or Resilience

A degree of stability is necessary for any ecosystem to continue to maintain itself. Some scientists have proposed that resilience to both external and internal changes (with the ability to rapidly return to its baseline functional profile; see above) is a key feature of a healthy gut microbiome (Bäckhed et al., 2012). This requires complex approaches to measurement and should be a topic of future investigation.

In individuals without disease, "health-associated" microbiota is preferred to the term "healthy microbiota," since gut microbiota composition alone cannot predict any state of health or disease according to currently available research. It may turn out that many possible states of gut microbiota are associated with health or indeed that a "dynamic equilibrium" better describes the gut microbiome of those in good health (Lloyd-Price et al., 2016). To add further complication, others have pointed out that "the gut microbiota of a healthy person may not be equivalent to a healthy microbiota. It is possible that the Western microbiota is actually dysbiotic and predisposes individuals to a variety of diseases" (Sonnenburg and Sonnenburg, 2014). Distinguishing between (1) an optimal microbiota, (2) one that may signal disease risk and (3) one that actively causes disease is a considerably challenging task that lies ahead.

REFERENCES

Abt, M.C., McKenney, P.T., Pamer, E.G., 2016. Clostridium difficile colitis: pathogenesis and host defence. Nat. Publ. Group 14 (10), 609–620.

Adami, A.J., Bracken, S.J., 2016. Breathing better through bugs: asthma and the microbiome. Yale J. Biol. Med. 89 (3), 309–324.

Ajouz, H., Mukherji, D., Shamseddine, A., 2014. Secondary bile acids: an underrecognized cause of colon cancer. World J. Surg. Oncol. 12, 164. Available from: http://www.ncbi.nlm.nih.gov/pubmed/24884764.

Almeida, R., Gerbaba, T., Petrof, E.O., 2016. Recurrent Clostridium difficile infection and the microbiome. J. Gastroenterol. 51 (1), 1–10.

Arpaia, N., et al., 2013. Metabolites produced by commensal bacteria promote peripheral regulatory T-cell generation. Nature 504 (7480), 451–455.

Arrieta, M.-C., et al., 2015. Early infancy microbial and metabolic alterations affect risk of childhood asthma. Sci. Transl. Med. 7 (307), 307ra152.

Ausubel, F., 2005. Are innate immune signaling pathways in plants and animals conserved? Nat. Immunol. 6 (105), 973–979.

Backert, S., et al., 2013. Transmigration route of Campylobacter jejuni across polarized intestinal epithelial cells: paracellular, transcellular or both? Cell Commun. Signal 11 (1), 72.

Bäckhed, F., et al., 2004. The gut microbiota as an environmental factor that regulates fat storage. Proc. Natl. Acad. Sci. U. S. A. 101 (44), 15718–15723. Available from: http://www.ncbi.nlm.nih.gov/pubmed/15505215.

Bäckhed, F., et al., 2012. Defining a healthy human gut microbiome: current concepts, future directions, and clinical applications. Cell Host Microbe 12 (5), 611–622. Available from: http://www.sciencedirect.com/science/article/pii/S1931312812003587.

Bajaj, J.S., et al., 2014. Altered profile of human gut microbiome is associated with cirrhosis and its complications. J. Hepatol. 60 (5), 940–947. Available from: http://www.ncbi.nlm.nih.gov/pubmed/24374295.

Baldridge, M.T., Turula, H., Wobus, C.E., 2016. Norovirus regulation by host and microbe. Trends Mol. Med. 22 (12), 1047–1059.

Balsari, A., et al., 1982. The fecal microbial population in the irritable bowel syndrome. Microbiologica 5 (3), 185–194. Available from: http://www.ncbi.nlm.nih.gov/pubmed/7121297.

Berer, K., et al., 2011. Commensal microbiota and myelin autoantigen cooperate to trigger autoimmune demyelination. Nature 479 (7374), 538–541. Available from: http://www.ncbi.nlm.nih.gov/pubmed/22031325.

Blanton, L.V., et al., 2016. Gut bacteria that prevent growth impairments transmitted by microbiota from malnourished children. Science 351 (6275), aad3311. Available from: http://www.ncbi.nlm.nih.gov/pubmed/26912898.

Bode, J.C., et al., 1984. Jejunal microflora in patients with chronic alcohol abuse. Hepato-Gastroenterology 31 (1), 30–34. Available from: http://www.ncbi.nlm.nih.gov/pubmed/6698486.

Bolton, D.J., 2015. Campylobacter virulence and survival factors. Food Microbiol. 48, 99–108.

Boursier, J., et al., 2016. The severity of nonalcoholic fatty liver disease is associated with gut dysbiosis and shift in the metabolic function of the gut microbiota. Hepatology 63 (3). Available from: http://www.ncbi.nlm.nih.gov/pubmed/26600078.

Brestoff, J.R., Artis, D., 2013. Commensal bacteria at the interface of host metabolism and the immune system. Nat. Immunol. 14 (7), 676–684.

Buffington, S.A., et al., 2016. Microbial reconstitution reverses maternal diet-induced social and synaptic deficits in offspring. Cell 165 (7), 1762–1775. Available From: http://www.ncbi.nlm.nih.gov/pubmed/27315483.

Caminero, A., et al., 2016. Duodenal bacteria from patients with Celiac disease and healthy subjects distinctly affect gluten breakdown and immunogenicity. Gastroenterology 151 (4), 670–683. Available from: http://linkinghub.elsevier.com/retrieve/pii/S0016508516347138.

Cani, P.D., et al., 2007. Metabolic endotoxemia initiates obesity and insulin resistance. Diabetes 56 (7). Available from: http://diabetes.diabetesjournals.org/content/56/7/1761.

Caricilli, A.M., Saad, M.J.A., 2013. The role of gut microbiota on insulin resistance. Nutrients 5 (3), 829–851. Available from: http://www.ncbi.nlm.nih.gov/pubmed/23482058.

Carter, G.P., et al., 2015. Defining the roles of TcdA and TcdB in localized gastrointestinal disease, systemic organ damage, and the host response during Clostridium difficile infections. MBio 6 (3), 1–10.

Cartwright, E.J., et al., 2013. Listeriosis outbreaks and associated food vehicles, United States, 1998-2008. Emerg. Infect. Dis. 19 (1), 1–9.

Chen, Y., et al., 2011. Characterization of fecal microbial communities in patients with liver cirrhosis. Hepatology 54 (2), 562–572. Available from: http://www.ncbi.nlm.nih.gov/pubmed/21574172.

Chen, J., et al., 2016. An expansion of rare lineage intestinal microbes characterizes rheumatoid arthritis. Genome Med. 8 (1), 43. Available from: http://genomemedicine.biomedcentral.com/articles/10.1186/s13073-016-0299-7.

Chung, H.K.L., et al., 2016. Genome analysis of Campylobacter concisus strains from patients with inflammatory bowel disease and gastroenteritis provides new insights into pathogenicity. Sci. Rep. 6, 38442.

Cohen, M.L., Tauxe, R.V., 1986. Drug-resistant salmonella in United States: perspective. Science 234, 964–969.

Corrêa-Oliveira, R., et al., 2016. Regulation of immune cell function by short-chain fatty acids. Clin. Transl. Immunol. 5 (4), e73. Available from: http://www.nature.com/doifinder/10.1038/cti.2016.17.

Dao, M.C., et al., 2016. Akkermansia muciniphila and improved metabolic health during a dietary intervention in obesity: relationship with gut microbiome richness and ecology. Gut 65 (3), 426–436. Available from: http://www.ncbi.nlm.nih.gov/pubmed/26100928.

Davies, J., Davies, D., 2010. Origins and evolution of antibiotic resistance. Microbiol. Mol. Biol. Rev. 74 (3), 417–433.

de Goffau, M.C., et al., 2013. Fecal microbiota composition differs between children with β-cell autoimmunity and those without. Diabetes 62 (4).

de Graaf, M., Villabruna, N., Koopmans, M.P., 2017. Capturing norovirus transmission. Curr. Opin. Virol. 22, 64–70.

Del Chierico, F., et al., 2014. Meta-omic platforms to assist in the understanding of NAFLD gut microbiota alterations: tools and applications. Int. J. Mol. Sci. 15 (1), 684–711. Available from: http://www.ncbi.nlm.nih.gov/pubmed/24402126.

Delzenne, N.M., Kok, N., 1998. Effect of non-digestible fermentable carbohydrates on hepatic fatty acid metabolism. Biochem. Soc. Trans. 26 (2), 228–230. Available from: http://www.ncbi.nlm.nih.gov/pubmed/9649752.

Distrutti, E., et al., 2016. Gut microbiota role in irritable bowel syndrome: new therapeutic strategies. World J. Gastroenterol. 22 (7), 2219–2241. Available from: http://www.ncbi.nlm.nih.gov/pubmed/26900286.

Etienne-Mesmin, L., et al., 2011. Interactions with M cells and macrophages as key steps in the pathogenesis of enterohemorragic Escherichia coli infections. PLoS One 6 (8).

Everard, A., et al., 2013. Cross-talk between Akkermansia muciniphila and intestinal epithelium controls diet-induced obesity. Proc. Natl. Acad. Sci. U. S. A. 110 (22), 9066–9071. Available from: http://www.ncbi.nlm.nih.gov/pubmed/23671105.

Fu, J., et al., 2015. The gut microbiome contributes to a substantial proportion of the variation in blood lipids. Circ. Res. 117, 817–824.

Furusawa, Y., et al., 2013. Commensal microbe-derived butyrate induces the differentiation of colonic regulatory T cells. Nature 504 (7480), 446–450.

Gil, F., et al., 2017. Molecular biology and genetics of anaerobes updates on Clostridium difficile spore biology. Anaerobe, 1–7.

Giloteaux, L., et al., 2016. Reduced diversity and altered composition of the gut microbiome in individuals with myalgic encephalomyelitis/chronic fatigue syndrome. Microbiome 4 (1), 30. Available from: http://microbiomejournal.biomedcentral.com/articles/10.1186/s40168-016-0171-4.

Guo, X., et al., 2015. Innate lymphoid cells control early colonization resistance against intestinal pathogens through ID2-dependent regulation of the microbiota. Immunity 42 (4), 731–743. Available from: http://www.ncbi.nlm.nih.gov/pubmed/25902484.

Heintz-Buschart, A., et al., 2016. Integrated multi-omics of the human gut microbiome in a case study of familial type 1 diabetes. Nat. Microbiol. 2, 16180. Available from: http://www.nature.com/articles/nmicrobiol2016180.

Hooper, L.V., Macpherson, A.J., 2010. Immune adaptations that maintain homeostasis with the intestinal microbiota. Nat. Rev. Immunol. 10 (3), 159–169.

Hsiao, E.Y., et al., 2013. Microbiota modulate behavioral and physiological abnormalities associated with neurodevelopmental disorders. Cell 155 (7), 1451–1463. Available from: http://www.ncbi.nlm.nih.gov/pubmed/24315484.

Ianiro, G., Tilg, H., Gasbarrini, A., 2016. Antibiotics as deep modulators of gut microbiota: between good and evil. Gut 65 (11), 1906–1915. Available from: http://gut.bmj.com/lookup/doi/10.1136/gutjnl-2016-312297.

Jiang, H., et al., 2015. Altered fecal microbiota composition in patients with major depressive disorder. Brain Behav. Immun. 48, 186–194. Available from: http://www.sciencedirect.com/science/article/pii/S0889159115001105.

Jones, M.K., et al., 2014. Enteric bacteria promote human and mouse norovirus infection of B cells. Science 346 (6210), 755–759.

Kang, D.J., et al., 2016. Gut microbiota drive the development of neuroinflammatory response in cirrhosis in mice. Hepatology 64 (4), 1232–1248. Available from: http://doi.wiley.com/10.1002/hep.28696.

Karlsson, F.H., et al., 2012. Symptomatic atherosclerosis is associated with an altered gut metagenome. Nat. Commun. 3, 1245. Available from: http://www.nature.com/doifinder/10.1038/ncomms2266.

Kelly, C.R., et al., 2016. Effect of fecal microbiota transplantation on recurrence in multiply recurrent clostridium difficile infection. Ann. Intern. Med. 165 (9), 609. Available from: http://annals.org/article.aspx?doi=10.7326/M16-0271.

Khan, M.J., et al., 2016. Role of gut microbiota in the aetiology of obesity: proposed mechanisms and review of the literature. J. Obes. 2016, 7353642. Available from: http://www.ncbi.nlm.nih.gov/pubmed/27703805.

Kimura, I., et al., 2014. The SCFA receptor GPR43 and energy metabolism. Front. Endocrinol. 5, 3–5.

Kirk, M.D., et al., 2015. World Health Organization estimates of the global and regional disease burden of 22 foodborne bacterial, protozoal, and viral diseases, 2010: a data synthesis. PLOS Med. 12 (12), e1001940.

Knip, M., Siljander, H., 2016. The role of the intestinal microbiota in type 1 diabetes mellitus. Nat. Rev. Endocrinol. 12 (3), 154–167. Available from: http://www.ncbi.nlm.nih.gov/pubmed/26729037.

Koeth, R.A., et al., 2013. Intestinal microbiota metabolism of L-carnitine, a nutrient in red meat, promotes atherosclerosis. Nat. Med. 19 (5), 576–585. Available from: http://www.ncbi.nlm.nih.gov/pubmed/23563705.

Kostic, A.D., et al., 2015. The dynamics of the human infant gut microbiome in development and in progression toward type 1 diabetes. Cell Host Microbe 17 (2), 260–273. Available from: http://www.ncbi.nlm.nih.gov/pubmed/25662751.

Lamas, B., et al., 2016. CARD9 impacts colitis by altering gut microbiota metabolism of tryptophan into aryl hydrocarbon receptor ligands. Nat. Med. 22 (6), 598–605. Available from: http://www.nature.com/doifinder/10.1038/nm.4102.

Laureto, L.M.O., Cianciaruso, M.V., Samia, D.S.M., 2015. Functional diversity: an overview of its history and applicability. Nat. Conservação 13 (2), 112–116. Available from: http://www.sciencedirect.com/science/article/pii/S1679007315000390.

Le Chatelier, E., et al., 2013. Richness of human gut microbiome correlates with metabolic markers. Nature 500 (7464), 541–546. Available from: http://www.ncbi.nlm.nih.gov/pubmed/23985870.

Lessa, F.C., et al., 2015. Burden of Clostridium difficile infection in the United States. N. Engl. J. Med. 372 (9), 825–834.

Li, J., et al., 2017. Gut microbiota dysbiosis contributes to the development of hypertension. Microbiome 5 (1), 14. Available from: http://microbiomejournal.biomedcentral.com/articles/10.1186/s40168-016-0222-x.

Liu, T.-X., Niu, H.-T., Zhang, S.-Y., 2015. Intestinal microbiota metabolism and atherosclerosis. Chin. Med. J. 128 (20), 2805–2811. Available from: http://www.ncbi.nlm.nih.gov/pubmed/26481750.

Lloyd-Price, J., Abu-Ali, G., Huttenhower, C., 2016. The healthy human microbiome. Genome Med. 8 (1), 51. Available from: http://genomemedicine.biomedcentral.com/articles/10.1186/s13073-016-0307-y.

López, P., et al., 2016. Th17 responses and natural IgM antibodies are related to gut microbiota composition in systemic lupus erythematosus patients. Sci. Rep. 6, 24072. Available from: http://www.nature.com/articles/srep24072.

Luna, R.A., et al., 2017. Distinct microbiome-neuroimmune signatures correlate with functional abdominal pain in children with autism spectrum disorder. Cell. Mol. Gastroenterol. Hepatol. 3 (2), 218–230. Available from: http://linkinghub.elsevier.com/retrieve/pii/S2352345X16301369.

Malik-Kale, P., Winfree, S., Steele-Mortimer, O., 2012. The bimodal lifestyle of intracellular Salmonella in epithelial cells: replication in the cytosol obscures defects in vacuolar replication. PLoS One 7 (6), 1–10.

Mechanick, J.I., Hurley, D.L., Garvey, W.T., 2016. Adiposity-based chronic disease as a new diagnostic term: american association of clinical endocrinologists and the american college of endocrinology position statement. Endocr. Pract. EP161688.PS, Available from: http://journals.aace.com/doi/10.4158/EP161688.PS.

Miyake, S., et al., 2015. Dysbiosis in the gut microbiota of patients with multiple sclerosis, with a striking depletion of species belonging to clostridia XIVa and IV clusters. In: Wilson, B.A. (Ed.), PLoS One 10 (9), e0137429. Available from: http://dx.plos.org/10.1371/journal.pone.0137429.

Monaco, C.L., et al., 2016. Altered virome and bacterial microbiome in human immunodeficiency virus-associated acquired immunodeficiency syndrome. Cell Host Microbe 19 (3), 311–322. Available from: http://www.ncbi.nlm.nih.gov/pubmed/26962942.

Mouzaki, M., et al., 2013. Intestinal microbiota in patients with nonalcoholic fatty liver disease. Hepatology 58 (1), 120–127. Available from: http://doi.wiley.com/10.1002/hep.26319.

Mutlu, E.A., et al., 2012. Colonic microbiome is altered in alcoholism. Am. J. Physiol. Gastrointest. Liver Physiol. 302 (9), G966–G978. Available from: http://www.ncbi.nlm.nih.gov/pubmed/22241860.

Nieto, P.A., et al., 2015. New insights about excisable pathogenicity islands in Salmonella and their contribution to virulence. Microbes Infect. 18, 302–309.

Ochoa-Reparáz, J., et al., 2010. A polysaccharide from the human commensal Bacteroides fragilis protects against CNS demyelinating disease. Mucosal Immunol. 3 (5), 487–495.

O'Connor, A., O'Morain, C.A., Ford, A.C., 2017. Population screening and treatment of Helicobacter pylori infection. Nat. Rev. Gastroenterol. Hepatol. Available from: http://www.ncbi.nlm.nih.gov/pubmed/28053340.

O'Keefe, S.J.D., et al., 2015. Fat, fibre and cancer risk in African Americans and rural Africans. Nat. Commun. 6, 6342. Available from: http://www.nature.com/doifinder/10.1038/ncomms7342.

Olesen, S.W., Alm, E.J., 2016. Dysbiosis is not an answer. Nat. Microbiol. 1, 16228. Available from: http://www.nature.com/articles/nmicrobiol2016228.

Ott, S.J., et al., 2004. Reduction in diversity of the colonic mucosa associated bacterial microflora in patients with active inflammatory bowel disease. Gut 53 (5), 685–693. Available from: http://www.ncbi.nlm.nih.gov/pubmed/15082587.

Ottman, N., et al., 2017. Pili-like proteins of Akkermansia muciniphila modulate host immune responses and gut barrier function. In: Sanz, Y. (Ed.), PLoS One 12 (3), e0173004. Available from: http://dx.plos.org/10.1371/journal.pone.0173004.

Pammi, M., et al., 2017. Intestinal dysbiosis in preterm infants preceding necrotizing enterocolitis: a systematic review and meta-analysis. Microbiome 5 (1), 31. Available from: http://www.ncbi.nlm.nih.gov/pubmed/28274256.

Pascal, V., et al., 2017. A microbial signature for Crohn's disease. Gut gutjnl–2016-313235. Available from: http://www.ncbi.nlm.nih.gov/pubmed/28179361.

Petersen, C., et al., 2014. Defining dysbiosis and its influence on host immunity and disease. Cell. Microbiol. 16 (7), 1024–1033. Available from: http://doi.wiley.com/10.1111/cmi.12308.

Petrof, E.O., et al., 2013. Stool substitute transplant therapy for the eradication of Clostridium difficile infection: "RePOOPulating" the gut. Microbiome 1 (1), 3. Available from: http://microbiomejournal.biomedcentral.com/articles/10.1186/2049-2618-1-3.

Pizarro-cerda, J., Ku, A., 2012. Entry of Listeria monocytogenes in Mammalian. Cold Spring Harb. Perspect. Med., 1–18.

Plovier, H., et al., 2016. A purified membrane protein from Akkermansia muciniphila or the pasteurized bacterium improves metabolism in obese and diabetic mice. Nat. Med. 23 (1), 107–113. Available from: http://www.nature.com/doifinder/10.1038/nm.4236.

Prawitt, J., Caron, S., Staels, B., 2011. Bile acid metabolism and the pathogenesis of type 2 diabetes. Curr. Diab. Rep. 11 (3), 160–166. Available from: http://www.ncbi.nlm.nih.gov/pubmed/21431855.

Qin, J., et al., 2012. A metagenome-wide association study of gut microbiota in type 2 diabetes. Nature 490 (7418), 55–60. Available from: http://www.nature.com/doifinder/10.1038/nature11450.

Qin, N., et al., 2014. Alterations of the human gut microbiome in liver cirrhosis. Nature 513 (7516), 59–64. Available from: http://www.ncbi.nlm.nih.gov/pubmed/25079328.

Raman, M., et al., 2013. Fecal microbiome and volatile organic compound metabolome in obese humans with nonalcoholic fatty liver disease. Clin. Gastroenterol. Hepatol. 11 (7), 868–875.e3. Available from: http://www.ncbi.nlm.nih.gov/pubmed/23454028.

Ridaura, V.K., et al., 2013. Gut microbiota from twins discordant for obesity modulate metabolism in mice. Science 341 (6150).

Rivas, M.A., et al., 2011. Deep resequencing of GWAS loci identifies independent rare variants associated with inflammatory bowel disease. Nat. Genet. 43, 1066–1073.

Roager, H.M., et al., 2016. Colonic transit time is related to bacterial metabolism and mucosal turnover in the gut. Nat. Microbiol. 1 (9), 16093. Available from: http://www.nature.com/articles/nmicrobiol201693.

Rook, G.A.W., Brunet, L.R., 2005. Microbes, immunoregulation, and the gut. Gut 54 (3), 317–320. Available from: http://www.ncbi.nlm.nih.gov/pubmed/15710972.

Round, J.L., Mazmanian, S.K., 2010. Inducible Foxp3+ regulatory T-cell development by\ na commensal bacterium of the intestinal microbiota. Proc. Natl. Acad. Sci. U. S. A. 107 (27), 12204–12209.

Sabat, R., Ouyang, W., Wolk, K., 2014. Therapeutic opportunities of the IL-22-IL-22R1 system. Nat. Rev. Drug Discov. 13 (1), 21–38.

Sadiq, S.M., et al., 2014. EHEC genomics: past, present, and future. Microbiol. Spectr., 1–13.

Sampson, T.R., et al., 2016. Gut microbiota regulate motor deficits and neuroinflammation in a model of Parkinson's disease. Cell 167 (6), 1469–1480.e12. Available from: http://www.ncbi.nlm.nih.gov/pubmed/27912057.

Sartor, R.B., Wu, G.D., 2017. Roles for intestinal bacteria, viruses, and fungi in pathogenesis of inflammatory bowel diseases and therapeutic approaches. Gastroenterology 152 (2), 327–339.e4. Available from: http://linkinghub.elsevier.com/retrieve/pii/S0016508516352350.

Scallan, E., et al., 2011. Foodborne illness acquired in the United States—major pathogens. Emerg. Infect. Dis. 17 (1), 7–15. Available from: http://wwwnc.cdc.gov/eid/article/17/1/P1-1101_article.htm.

Scanlan, E., et al., 2017. Relaxation of DNA supercoiling leads to increased invasion of epithelial cells and protein secretion by Campylobacter jejuni. Mol. Microbiol. 104 (February), 92–104.

Scheperjans, F., et al., 2015. Gut microbiota are related to Parkinson's disease and clinical phenotype. Mov. Disord. 30 (3), 350–358. Available from: http://www.ncbi.nlm.nih.gov/pubmed/25476529.

Schwarzer, M., et al., 2016. Lactobacillus plantarum strain maintains growth of infant mice during chronic undernutrition. Science 351 (6275). Available from: http://science.sciencemag.org/content/351/6275/854.

Shawcross, D.L., et al., 2010. Ammonia and the neutrophil in the pathogenesis of hepatic encephalopathy in cirrhosis. Hepatology 51 (3), 1062–1069. Available from: http://www.ncbi.nlm.nih.gov/pubmed/19890967.

Smith, P., Howitt, M., Panikov, N., Michaud, M., Gallini, C., Bohlooly, Y.M., Glickman, J.G.W., 2013. The microbial metabolites, short-chain fatty acids, regulate colonic treg cell homeostasis. Science 341, 569–574.

Sokol, H., et al., 2016. Fungal microbiota dysbiosis in IBD. Gut gutjnl–2015-310746. Available from: http://www.ncbi.nlm.nih.gov/pubmed/26843508.

Sonnenberg, G.F., et al., 2012. Innate lymphoid cells promote anatomical containment of lymphoid-resident commensal bacteria. Science 336, 1321–1325.

Sonnenburg, E.D., Sonnenburg, J.L., 2014. Starving our microbial self: the deleterious consequences of a diet deficient in microbiota-accessible carbohydrates. Cell Metab. 20 (5), 779–786.

Strachan, D.P., 2000. Family size, infection and atopy: the first decade of the "hygiene hypothesis". Thorax 55 (Suppl. 1), S2–S10.

Subramanian, S., et al., 2014. Persistent gut microbiota immaturity in malnourished Bangladeshi children. Nature 510 (7505), 417–421. Available from: http://www.ncbi.nlm.nih.gov/pubmed/24896187.

Swiatczak, B., Cohen, I.R., 2015. Gut feelings of safety: tolerance to the microbiota mediated by innate immune receptors. Microbiol. Immunol. 59 (10), 573–585.

Szabo, G., 2015. Gut-liver axis in alcoholic liver disease. Gastroenterology 148 (1), 30–36. Available from: http://www.ncbi.nlm.nih.gov/pubmed/25447847.

Sze, M.A., Schloss, P.D., 2016. Looking for a signal in the noise: revisiting obesity and the microbiome. MBio 7 (4), e01018–16. Available from: http://www.ncbi.nlm.nih.gov/pubmed/27555308.

Tang, W.H.W., et al., 2013. Intestinal microbial metabolism of phosphatidylcholine and cardiovascular risk. N. Engl. J. Med. 368 (17), 1575–1584. Available from: http://www.ncbi.nlm.nih.gov/pubmed/23614584.

Tap, J., et al., 2017. Identification of an intestinal microbiota signature associated with severity of irritable bowel syndrome. Gastroenterology 152 (1), 111–123.e8. Available from: http://linkinghub.elsevier.com/retrieve/pii/S0016508516351745.

Thabane, M., Marshall, J.K., 2009. Post-infectious irritable bowel syndrome. World J. Gastroenterol. 15 (29), 3591–3596. Available from: http://www.ncbi.nlm.nih.gov/pubmed/19653335.

Thaiss, C.A., et al., 2016. The microbiome and innate immunity. Nature 535 (7610), 65–74.

Theriot, C.M., Bowman, A., Young, V.B., 2015. Antibiotic-induced alterations of the gut microbiota alter secondary bile acid production and allow for clostridium difficile spore germination and outgrowth in the large intestine. mSphere 1 (1), e00045–15.

Tilg, H., Cani, P.D., Mayer, E.A., 2016. Gut microbiome and liver diseases. Gut 65 (12), 2035–2044. Available from: http://gut.bmj.com/lookup/doi/10.1136/gutjnl-2016-312729.

Touati, E., 2010. When bacteria become mutagenic and carcinogenic: lessons from H. pylori. Mutat. Res. Genet. Toxicol. Environ. Mutagen. 703 (1), 66–70. Available from: http://www.ncbi.nlm.nih.gov/pubmed/20709622.

Tremlett, H., et al., 2017. The gut microbiome in human neurological disease: a review. Ann. Neurol. 81 (3), 369–382. Available from: http://www.ncbi.nlm.nih.gov/pubmed/28220542.

Turnbaugh, P.J., et al., 2006. An obesity-associated gut microbiome with increased capacity for energy harvest. Nature 444 (7122), 1027–1131. Available from: http://www.nature.com/doifinder/10.1038/nature05414.

Turnbaugh, P.J., et al., 2009. A core gut microbiome in obese and lean twins. Nature 457 (7228), 480–484. Available from: http://www.ncbi.nlm.nih.gov/pubmed/19043404.

van Nood, E., et al., 2013. Duodenal infusion of donor feces for recurrent *Clostridium difficile*. N. Engl. J. Med. 368 (5), 407–415. Available from: http://www.nejm.org/doi/abs/10.1056/NEJMoa1205037.

Verhoef, L., et al., 2015. Norovirus genotype profiles associated with foodborne transmission. Emerg. Infect. Dis. 21 (4), 592–599.

Vindigni, S.M., Surawicz, C.M., 2015. C. difficile infection: changing epidemiology and management paradigms. Clin. Transl. Gastroenterol. 6 (7), e99.

Wacklin, P., et al., 2014. Altered duodenal microbiota composition in celiac disease patients suffering from persistent symptoms on a long-term gluten-free diet. Am. J. Gastroenterol. 109 (12), 1933–1941. Available from: http://www.ncbi.nlm.nih.gov/pubmed/25403367.

Walker, A.W., et al., 2011. High-throughput clone library analysis of the mucosa-associated microbiota reveals dysbiosis and differences between inflamed and non-inflamed regions of the intestine in inflammatory bowel disease. BMC Microbiol. 11 (1), 7. Available from: http://bmcmicrobiol.biomedcentral.com/articles/10.1186/1471-2180-11-7.

Walters, W.A., Xu, Z., Knight, R., 2014. Meta-analyses of human gut microbes associated with obesity and IBD. FEBS Lett. 588 (22), 4223–4233. Available from: http://doi.wiley.com/10.1016/j.febslet.2014.09.039.

Wang, Z., et al., 2014. Prognostic value of choline and betaine depends on intestinal microbiota-generated metabolite trimethylamine-N-oxide. Eur. Heart J. 35 (14), 904–910. Available from: http://www.ncbi.nlm.nih.gov/pubmed/24497336.

WHO, 2017. WHO | The top 10 causes of death. WHO. Available from: http://www.who.int/mediacentre/factsheets/fs310/en/.

Williams, B., Landay, A., Presti, R.M., 2016. Microbiome alterations in HIV infection a review. Cell. Microbiol. 18 (5), 645–651. Available from: http://doi.wiley.com/10.1111/cmi.12588.

Wright, E.K., et al., 2015. Recent advances in characterizing the gastrointestinal microbiome in Crohn's disease: a systematic review. Inflamm. Bowel Dis. 21 (6), 1219–1228. Available from: http://www.ncbi.nlm.nih.gov/pubmed/25844959.
Wu, H., Tremaroli, V., Bäckhed, F., 2015. Linking microbiota to human diseases: a systems biology perspective. Trends Endocrinol. Metab. 26 (12), 758–770. Available from: http://linkinghub.elsevier.com/retrieve/pii/S1043276015001940.
Zackular, J.P., et al., 2013. The gut microbiome modulates colon tumorigenesis. MBio 4 (6), e00692–13. Available from: http://www.ncbi.nlm.nih.gov/pubmed/24194538.
Zhang, J., et al., 2016. Evolution and diversity of Listeria monocytogenes from clinical and food samples in Shanghai, China. Front. Microbiol. 7, 1–9.
Zhuang, X., et al., 2017. Alterations of gut microbiota in patients with irritable bowel syndrome: a systematic review and meta-analysis. J. Gastroenterol. Hepatol. 32 (1), 28–38. Available from: http://doi.wiley.com/10.1111/jgh.13471.
Zupancic, M.L., et al., 2012. Analysis of the gut microbiota in the old order Amish and its relation to the metabolic syndrome. In: Thameem, F. (Ed.), PLoS One 7 (8), e43052. Available from: http://dx.plos.org/10.1371/journal.pone.0043052.

CHAPTER 5
Genetic and Environmental Influences on Gut Microbiota

Objectives

- To gain perspective on the range of factors known to account for variability in gut microbiota composition from person to person.
- To learn what has been discovered about the heritability of bacterial taxa in the gut.
- To understand the extent to which environmental factors—including medications, geography, living environment, infections, fitness, stress, and sleep—influence gut microbiota composition and/or function.

The gut microbiome of humans is characterized by significant variability from person to person (Eckburg et al., 2005), with both genetic and environmental factors contributing to these individual variations. Fig. 5.1 shows the key factors, described in this chapter, known to influence the human gut microbiome. A large-cohort study in a Dutch population found features of gut microbiota composition that correlated with both intrinsic factors, such as stool consistency and fecal chromogranin A (which is an important disease biomarker), and exogenous factors—primarily diet and medications (Zhernakova et al., 2016). In this study, 126 measured factors explained 18.7% of the observed variation in gut microbiome composition between individuals. A similar population-level analysis in both Belgian and Dutch individuals found, of all the measured environmental factors, medication explained most of the variation in gut microbial composition; this was followed by diet and then by other lifestyle factors (Falony et al., 2016). While researchers do not yet know everything that contributes to an individual's unique gut microbiome at a single point in time, they are beginning to uncover some of the most important genetic and environmental contributors.

GENETIC INFLUENCES

Studies on twin pairs have uncovered the names of some heritable taxa in the gut microbial community. Work led by Ruth Ley found, when analyzing the fecal samples of 416 twin pairs from the TwinsUK cohort, that

Gut Microbiota
https://doi.org/10.1016/B978-0-12-810541-2.00005-1

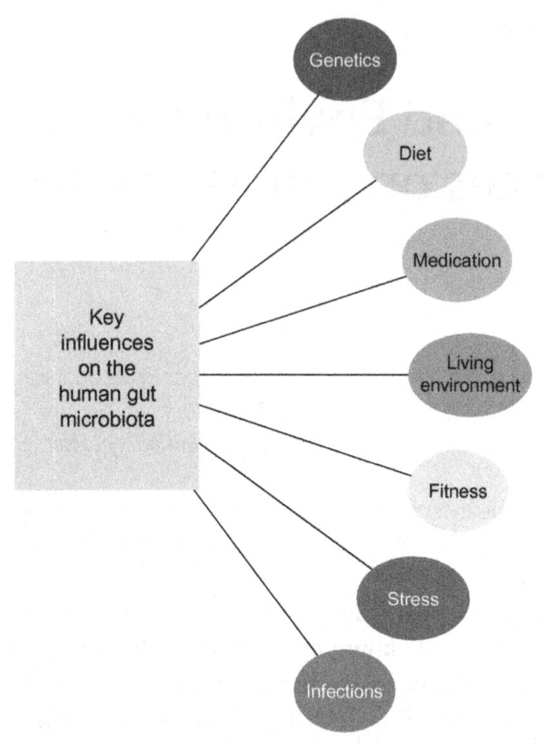

Fig. 5.1 An overview of the known factors that influence human gut microbiota composition.

human genetics particularly influenced the abundance of bacteria in the family Christensenellaceae; these bacteria co-occurred with other heritable bacteria and with methanogenic archaea. Moreover, Christensenellaceae and its group of co-occurring microorganisms were associated with a lower body mass index. When a human-obesity-associated microbiota and *Christensenella minuta* were transferred to germ-free mice, the consortia altered the gut microbiota and reduced weight gain in the animals, providing evidence that the gut bacteria influenced by human genes may impact host metabolism and/or body weight (Goodrich et al., 2014).

Follow-up work in 2016 that included more than twice as many twin pairs ($n = 1126$) found heritable bacteria to be stable over time, as might be expected. The researchers noted additional heritable bacterial taxa in the gut and found associations between these heritable taxa and genes related to diet, metabolism, and olfaction, as well as immune barrier defense and self-recognition/nonself-recognition (Goodrich et al., 2016).

As a specific example related to disease, other researchers found the colonic microbiota of mice and humans that were genetically abnormal fucosyltransferase 2 nonsecretors was altered at both compositional and functional levels; they speculated that this human gene-microbiota association could explain the connection of the genotype with increased susceptibility to Crohn's disease (Tong et al., 2014).

The correlations between genes and bacteria in the gut appear to be no accident, from an evolutionary perspective. Moeller and colleagues did a comparison on the evolutionary origins of bacterial lineages and showed that multiple lineages of the dominant bacterial groups in the human gut arose through cospeciation with humans, chimpanzees, bonobos, and gorillas; this occurred over a time scale of approximately 15 million years (Moeller et al., 2016). During hominid evolution, nuclear, mitochondrial, and gut bacterial genomes appeared to have diversified at the same time—and possibly, say the researchers, to assist in shaping the hominid immune system and development.

ENVIRONMENTAL INFLUENCES

Data show several major environmental factors—that is, potentially modifiable factors—that influence gut microbiota composition. Medication and diet stand out as the two primary environmental factors of influence, according to the large-cohort studies from Europe mentioned above (Falony et al., 2016; Zhernakova et al., 2016) and parallel work examining mechanism. Diet will be discussed in detail in Chapter 6. The key medications that may play a role in determining gut microbiota composition from individual to individual, along with some additional factors, are discussed below.

Medications

Medications exert significant effects on gut microbiota composition. In the two largest human cohorts to date, drug intake explained the largest total variance in gut microbiota composition from person to person, accounting for 10.04% of microbiome composition (Falony et al., 2016). Falony et al. found 13 drugs significantly associated with gut microbiota composition, including various antibiotics, osmotic laxatives, treatments for inflammatory bowel disease (IBD), female hormones, benzodiazepines, antidepressants, and antihistamines; a similar study added to this list proton pump inhibitors (PPIs), metformin, and statins (Zhernakova et al., 2016). These associations are summarized in Table 5.1. It should be noted that these were

Table 5.1 Associations between various medication types and changes in the human gut microbiota

Medication type	Gut microbial community composition change	Richness
Antidepressants	X	↑
Antihistamines	X	
Benzodiazepines	X	
β-Lactam antibiotics	X	↓
Estrogens	X	
Immunosuppressants	X	
Mesalazine (IBD treatment)		↓
Metformin	X	
Osmotic laxatives	X	
Proton pump inhibitors	X	
Statins	X	

Based on Falony, G., et al., 2016. Population-level analysis of gut microbiome variation. Science 352 (6285); Zhernakova, A., Kurilshikov, A., Bonder, M., Tigchelaar, E., Schirmer, M., et al., 2016. Population-based metagenomics analysis reveals markers for gut microbiome composition and diversity. Science 352 (6285), 565–569.

correlational studies, and mechanistic evidence linking some of these drugs with the gut microbiome is nonexistent.

Complex interactions occur between gut microbiota and xenobiotics (foreign substances, including drugs); see Fig. 5.2 for various mechanisms of interaction between gut microbiota and medications (Spanogiannopoulos et al., 2016). Importantly, the use of particular medications appears to change associations between gut microbiota and other variables (Falony et al., 2016). This highlights both the interdependency of factors determining gut microbiota composition and the necessity of controlling for medication use in future clinical studies on gut microbiota.

Antibiotics

Of all medications, antibiotics are the most well-studied with respect to the gut microbiome. For the most part, antibiotics seem to induce temporary effects on gut microbiota composition in healthy adults. A study of 12 healthy volunteers, for example, found increases or decreases in specific gut bacterial taxa upon administration of antibiotics, with different antibiotics (in this case, linezolid vs amoxicillin/clavulanic acid) impacting different bacteria. Composition returned to normal after 35 days (Lode et al., 2001).

Evidence suggests, however, that the gut microbiota is not always perfectly resilient to antibiotic perturbation. One detailed analysis showed treatment with ciprofloxacin affected approximately a third of bacterial taxa

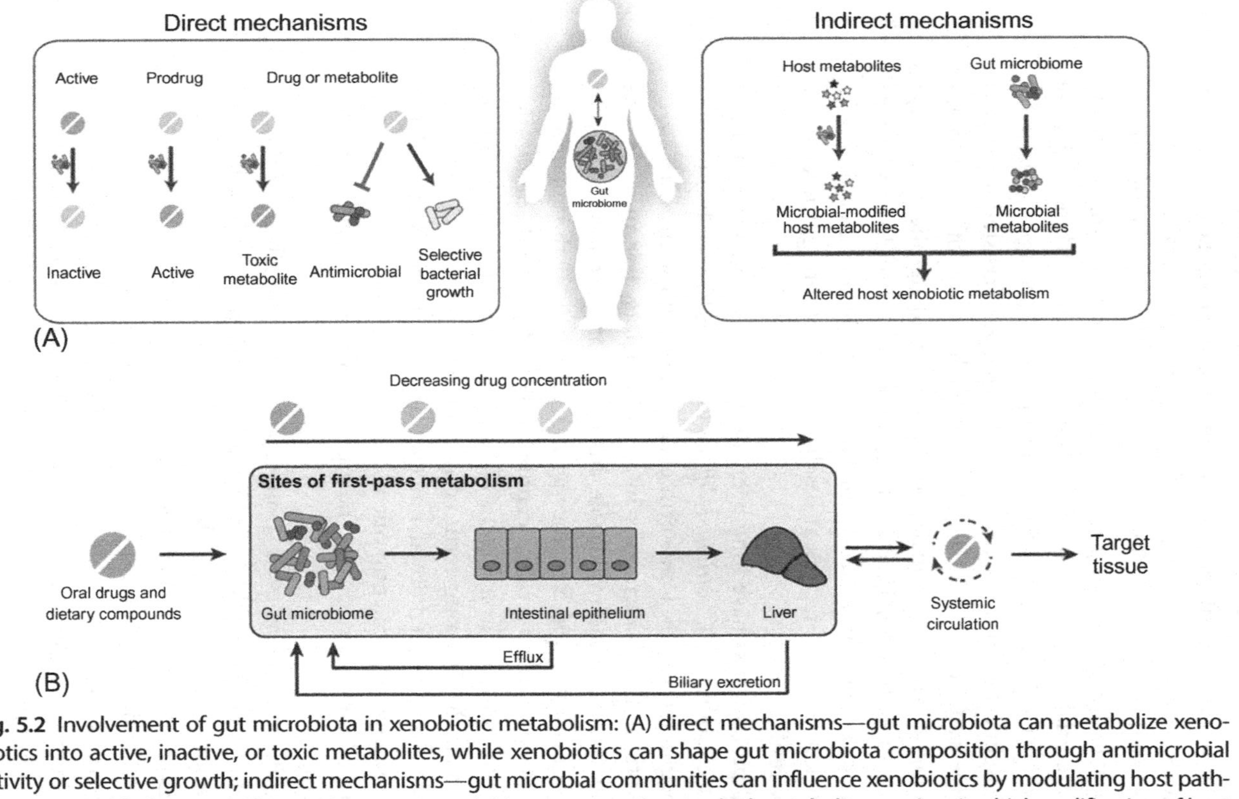

Fig. 5.2 Involvement of gut microbiota in xenobiotic metabolism: (A) direct mechanisms—gut microbiota can metabolize xenobiotics into active, inactive, or toxic metabolites, while xenobiotics can shape gut microbiota composition through antimicrobial activity or selective growth; indirect mechanisms—gut microbial communities can influence xenobiotics by modulating host pathways responsible for metabolism and transport, and this may occur via microbial metabolites or via microbial modification of host metabolites. (B) The gut microbiota is involved in **first-pass metabolism** of oral drugs, whereby concentration is reduced through metabolism in the intestine and liver before the compound reaches systemic circulation. Gut microbes may metabolize drugs at several stages: prior to absorption, following efflux from the gut epithelium, or after biliary excretion from the liver. *(Reproduced with permission from Spanogiannopoulos, P., Bess, E.N., Carmody, R.N., Turnbaugh, P.J., 2016. The microbial pharmacists within us: a metagenomic view of xenobiotic metabolism. Nat. Rev. Microbiol. 14, 273–287. Macmillan Publishers Ltd: Nature Reviews Microbiology. Copyright 2016.)*

in the stool samples of healthy humans, and despite individual differences, the researchers generally saw the taxonomic richness, diversity, and evenness decrease with the antibiotic treatment. The microbiota composition mostly bounced back by 4 weeks post treatment, but several taxa did not recover even within 6 months (Dethlefsen et al., 2008). In another investigation, six volunteers receiving oral amoxicillin showed a major shift in dominant bacterial species starting 24h after treatment; within 30 days of treatment, their fecal microbiota had recovered to an average similarity of 88% with baseline, but in one participant, the changes persisted for at least 2 months (De La Cochetière et al., 2005). A further study of eight volunteers receiving a 7-day clindamycin treatment found significant changes in fecal microbiota that persisted over the study period of 2 years; in particular, *Bacteroides* never returned to original levels (Jernberg et al., 2007). It is not yet known what factors at baseline may promote gut microbiota community resilience.

Medical professionals have long been aware that, in humans, antibiotics increase susceptibility to *Clostridium difficile* infection. Research on mouse models is beginning to link long-term changes in health status with short-term antibiotic-induced alterations in gut microbiota. For instance, one mouse study found that metronidazole, and to a greater extent vancomycin, led to a loss of colonization resistance against pathogens: with vancomycin, the mice showed a longer-term susceptibility to *C. difficile* and colonization by vancomycin-resistant *Enterococcus*, *Klebsiella pneumoniae*, and *Escherichia coli* (Lewis et al., 2015). Long-term consequences of antibiotic use in humans still need to be pinned down, but according to scientist Martin Blaser, any health consequences of antibiotic use are likely attributable to the "collateral damage" they exert on the normal microbiota of healthy humans (Blaser, 2016).

In early life, the case is even stronger for antibiotics affecting the gut microbiota in ways that could have lasting consequences for health. Many studies have found an association between antibiotic use in early life and immune-mediated diseases later on: for example, in a large cohort of American children, antibiotic exposure in the first 6 months was associated with increased risk of asthma by age 6 (Risnes et al., 2011). The theory that these links between antibiotics and health are mechanistically connected through the gut microbiota is gaining support.

Work from Finland has shown both gut microbiota shifts with early use of antibiotics and correlations with subsequent immune and metabolic diseases: Korpela and colleagues showed the use of macrolides (a group of antibiotics that are especially effective against Gram-positive bacteria like staphylococci and streptococci) in children aged 2–7 was associated with a sustained shift in gut microbiome composition and function, with the antibiotic-exposed children showing a decrease in Actinobacteria, an increase in Bacteroidetes and Proteobacteria, a decrease in bile salt hydrolase, and an increase in macrolide resistance genes. Penicillins, however, appeared to have weaker effects on the gut microbiota than macrolides. In children who had received more than two courses of macrolides before age 2, their antibiotic use was strongly correlated with greater body mass index and an increased risk of asthma (Korpela et al., 2016).

Rodent work lends further support to these early-life connections. A 2012 mouse study demonstrated that subtherapeutic antibiotics (i.e., those given at a lower dose than normally required for a therapeutic effect) altered gut microbiota composition, with a shift toward more colonic short-chain fatty acids (SCFAs) and an increase in microbial genes encoding SCFA production, and furthermore, these antibiotics led to increased metabolism-related hormones and adiposity (Cho et al., 2012). The researchers then showed in 2014 that low-dose penicillin in mice, limited to early life, induced a short-term gut microbiota perturbation but a long-term increase in body fat. The group established that the antibiotic-altered microbiota had a causal role in the metabolic changes by transferring the microbiota to germ-free mice and observing that it was sufficient to transfer the obese phenotype (Cox et al., 2014). While not yet proved in humans, this work highlights early life as a possibly crucial time for setting up host-microbe metabolic interactions.

Metformin

Multiple studies support the idea that metformin, a very common medication for management of type 2 diabetes, has distinct effects on the gut microbiome in humans. In a landmark study by Forslund and colleagues, which examined 784 individuals with type 2 diabetes, the researchers were able to untangle the gut microbiota signature of the drug from that of type 2 diabetes itself (Forslund et al., 2015). The microbiota appeared to mediate the therapeutic effects of metformin via SCFA production—the researchers' functional analysis revealed that the gut microbiota of metformin-treated individuals had a greater potential to produce the SCFAs butyrate and propionate. The lack of butyrate-producing taxa in the type 2 diabetes disease-associated microbiota was found to be partly ameliorated by metformin.

A large-cohort association study found those using metformin had an increased abundance of *E. coli* and changes in microbiota functions (Zhernakova et al., 2016). As for mechanistic support, in mice with high-fat diet-induced obesity, those treated with metformin showed improved serum glucose levels, body weight, and total cholesterol levels compared with those not treated with metformin. Both *Akkermansia muciniphila* (bacteria with therapeutic potential in metabolic disease; see Chapter 4) and *Clostridium cocleatum* increased after metformin treatment, and 18 Kyoto Encyclopedia of Genes and Genomes (KEGG) metabolic pathways, including those for sphingolipid and fatty acid metabolism, were upregulated (Lee and Ko, 2014).

Proton Pump Inhibitors (PPIs)

PPIs (e.g., omeprazole) are used to treat GI disorders like peptic ulcers and gastroesophageal reflux through suppression of gastric acid production. In a study of fecal samples of 1827 healthy twins, lower abundance of gut commensals and lower microbial diversity were found in those who used PPIs, along with increased abundance of oral and upper GI tract commensals. Streptococcaceae were significantly increased in the lower gut. These effects are attributed to the drug's removal of the low pH barrier between the upper GI tract and the lower gut, causing commensals normally killed by the acidic environment of the stomach to translocate into the lower GI tract (Jackson et al., 2016). Meanwhile, in a large cohort, PPI users ($n = 95$) showed changes in 33 bacterial pathways—most significantly the pathway of 2,3-butanediol biosynthesis, which affects the amount of acid produced during fermentation (Zhernakova et al., 2016). Scientists do not yet know if any of these gut microbiota associations relate to the increased risk of infections with PPI use.

GEOGRAPHY

Humans from different countries may have systematic differences in gut microbial composition, although variations in the processing of samples from lab to lab could confound these observations. Country-specific microbial signatures have been observed (Li et al., 2014); for instance, one Chinese study showed that the fecal microbiota of healthy young adults is clustered by ethnicity and/or geography rather than by lifestyle factors (Zhang et al., 2015). Observed geographical differences must be interpreted with caution given the available evidence, since dramatically different diets in various populations (De Filippo et al., 2010) and other confounding variables (including genetics) could possibly drive them.

In certain disorders, such as irritable bowel syndrome (Zhuang et al., 2017), the gut microbial signature associated with the disease may depend on the geographic location of the individuals studied. More research is warranted; caution should be exercised in attempts to identify gut microbial biomarkers of disease in the future, lest they differ by geography.

LIVING ENVIRONMENT

The aspects of the living environment that appear to influence gut microbiota composition include farm exposure, the presence of siblings, and a dog in the household.

A cross-sectional survey on children from rural Austria, Germany, and Switzerland found both exposure to stables and consumption of farm milk before age 1 were protective against asthma, hay fever, and atopic sensitization. Those who were exposed to stables until age 5 had the lowest frequencies of these conditions. Although the gut microbiota was not analyzed in these children, researchers cited "exposure to microbial compounds" as a possible factor involved in the mechanism (Riedler et al., 2001).

Then, in a study published in the *New England Journal of Medicine* (Stein et al., 2016), researchers attempted to uncover why US agricultural populations of Amish have significantly lower incidences of asthma and allergic sensitization than populations of Hutterites, even though the two groups have similar lifestyles and genetic ancestry. The microbial compositions of dust samples from Amish and Hutterite households significantly differed (with Amish household dust having significantly higher levels of bacterial endotoxins), and in conjunction, children from the two communities significantly differed in the proportions, phenotypes, and functions of certain innate immune cells: Amish children had increased levels of neutrophils and decreased levels of eosinophils, showing more robust innate immune systems. Also, in a mouse model of allergic asthma, dust extracts from Amish (but not Hutterite) homes affected immune responses and were protective against airway hyperresponsiveness. The researchers noted that Amish people follow traditional farming practices, while Hutterites use industrialized farming practices; some related aspect of the Amish environment may be protective against asthma by affecting innate immune responses, possibly through the gut microbiota.

Family or household membership also appears to influence the gut microbiota. When researchers studied the gut microbiota of eight healthy individuals of the same family (two parents and six children who ranged from 2 months to 10 years), its members could be distinguished from normal individuals from the same geographic region even though each person had a slightly different gut bacterial community. The children's gut microbiota compositions showed greater similarity with each other than those of the mother with the father, underlining the unique similarity between siblings. Curiously, the children's microbiota were equally similar to the mother and the father, despite the fact that they were homeschooled and shared an almost identical diet and environment with the mother only (Schloss et al., 2014). A different paper reported that, in early life, having older siblings correlated with an increased relative abundance of several bacterial taxa (*Haemophilus* and *Faecalibacterium* at 9 months, and *Barnesiella*, *Odoribacter*, *Asaccharobacter*, and *Gordonibacter* at 18 months) (Laursen et al., 2015), while a study of younger infants (4-month olds; $n=24$) found that microbiota richness and diversity were decreased in those with older siblings (Azad et al., 2013). While the occurrence of several allergic disorders is negatively correlated with number of siblings (Strachan et al., 1997), it is too early to know whether the gut microbiota is responsible for this association.

In humans, exposure to a dog early in life is protective against allergic disease. A mouse study made a possible mechanistic link between dog-associated house dust and health: mice exposed to dog-associated house dust showed a reduction in total number of airway T cells and a downregulation of Th2-related airway responses and of mucin secretion. These mice had a distinct composition of cecal microbiota, enriched for *Lactobacillus johnsonii* and other species. When the researchers supplemented other mice with *L. johnsonii*, the mice were protected during an airway allergen challenge. This work shows the inclusion of *L. johnsonii* among inhaled microbial exposures could be important for influencing adaptive immunity at remote mucosal surfaces in a manner that protects against respiratory insults (Fujimura et al., 2014).

INFECTIONS

It appears that infections can also disrupt the gut microbiota and that this is potentially connected with long-lasting effects on immune function. Work from the lab of Yasmine Belkaid showed an association between an infectious agent and chronic disease through the microbiota: after the resolution of *Yersinia pseudotuberculosis* infection in mice, researchers observed sustained inflammation and lymphatic leakage in the mesenteric adipose tissue that resulted in persistently compromised mucosal immune functions. Importantly, a microbiota was required for this inflammatory response to be maintained (Fonseca et al., 2015). A single acute infection was thus shown to initiate gut-microbiota-enabled "immunological scarring." Applicability of this research to humans remains to be seen.

FITNESS

A recent study showed that in healthy individuals, higher cardiorespiratory fitness (as measured by peak oxygen uptake) correlated with increases in both microbial diversity and fecal butyrate; the researchers also found a core set of functions rather than a core set of bacterial taxa in individuals with high fitness (Estaki et al., 2016). In another study, professional athletes (rugby players), who had a unique dietary pattern and a high level of physical activity, showed a higher-diversity gut microbiota compared with controls (Clarke et al., 2014). Although these data support the idea that physical fitness affects the gut microbiota in beneficial ways, microbes are not yet definitively connected with the known benefits of exercise for body and brain function.

STRESS

Psychological stress, whether acute or chronic, has been proposed as a factor that may modify the gut microbiota. "Top-down" influences on gut microbiota composition and function are relatively unknown in humans, but animal models have allowed some valuable insights. According to one hypothesis, since the brain normally contributes to the maintenance of the intestinal mucus layer and biofilm that provide a habitat for bacteria, psychological stressors or other brain factors could change this habitat, altering microbiota composition or total bacterial biomass (Carabotti et al., 2015). A preliminary study in pigs, for instance, found that a rush of noradrenaline increased pathogenic *E. coli* adherence to the intestinal mucosa (Chen et al., 2006). More generally, it is also possible that under the brain's direction, secretion of molecules by neurons, immune cells, and enterochromaffin cells can alter the gut microbiota (Carabotti et al., 2015).

SLEEP

Very preliminary work shows sleep deprivation may have an effect on the human gut microbiota. A study of young, normal-weight individuals revealed relatively subtle effects: The Firmicutes to Bacteroidetes ratio was affected by partial sleep deprivation, and a short sleep duration was associated with higher abundances of the families Coriobacteriaceae and Erysipelotrichaceae and a lower abundance of Tenericutes—microbiota characteristics with possible links to metabolic disorders (Benedict et al., 2016). But another study showed no effect of sleep restriction on gut microbiota composition in either rats or humans (Zhang et al., 2017). More is needed on the connections between sleep and gut microbiota composition and whether gut microbiota characteristics play a role in the metabolic alterations that often accompany sleep deprivation.

REFERENCES

Azad, M.B., et al., 2013. Infant gut microbiota and the hygiene hypothesis of allergic disease: impact of household pets and siblings on microbiota composition and diversity. Allergy Asthma Clin. Immunol. 9 (1), 15. Available from: http://www.ncbi.nlm.nih.gov/pubmed/23607879.

Benedict, C., et al., 2016. Gut microbiota and glucometabolic alterations in response to recurrent partial sleep deprivation in normal-weight young individuals. Mol. Metab. 5 (12), 1175–1186. Available from: http://linkinghub.elsevier.com/retrieve/pii/S2212877816301934.

Blaser, M.J., 2016. Antibiotic use and its consequences for the normal microbiome. Science 352 (6285), 544–545. Available from: http://www.sciencemag.org/cgi/doi/10.1126/science.aad9358.

Carabotti, M., et al., 2015. The gut-brain axis: interactions between enteric microbiota, central and enteric nervous systems. Ann. Gastroenterol. 28 (2), 203–209. Available from: http://www.ncbi.nlm.nih.gov/pubmed/25830558.

Chen, C., et al., 2006. Mucosally-directed adrenergic nerves and sympathomimetic drugs enhance non-intimate adherence of Escherichia coli O157:H7 to porcine cecum and colon. Eur. J. Pharmacol. 539 (1–2), 116–124. Available from: http://www.ncbi.nlm.nih.gov/pubmed/16687138.

Cho, I., et al., 2012. Antibiotics in early life alter the murine colonic microbiome and adiposity. Nature 488 (7413), 621–626. Available from: http://www.nature.com/doifinder/10.1038/nature11400.

Clarke, S.F., et al., 2014. Exercise and associated dietary extremes impact on gut microbial diversity. Gut 63 (12), 1913–1920. Available from: http://gut.bmj.com/lookup/doi/10.1136/gutjnl-2013-306541.

Cox, L.M., et al., 2014. Altering the intestinal microbiota during a critical developmental window has lasting metabolic consequences. Cell 158 (4), 705–721. Available from: http://www.ncbi.nlm.nih.gov/pubmed/25126780.

De Filippo, C., et al., 2010. Impact of diet in shaping gut microbiota revealed by a comparative study in children from Europe and rural Africa. Proc. Natl. Acad. Sci. U. S. A. 107 (33), 14691–14696.

De La Cochetière, M.F., et al., 2005. Resilience of the dominant human fecal microbiota upon short-course antibiotic challenge. J. Clin. Microbiol. 43 (11), 5588–5592. Available from: http://www.ncbi.nlm.nih.gov/pubmed/16272491.

Dethlefsen, L., et al., 2008. The pervasive effects of an antibiotic on the human gut microbiota, as revealed by deep 16S rRNA sequencing. In: Eisen, J.A. (Ed.), PLoS Biol. 6 (11), e280. Available from: http://dx.plos.org/10.1371/journal.pbio.0060280.

Eckburg, P.B., et al., 2005. Diversity of the human intestinal microbial flora. Science 308 (5728), 1635–1638. Available from: http://www.ncbi.nlm.nih.gov/pubmed/15831718.

Estaki, M., et al., 2016. Cardiorespiratory fitness as a predictor of intestinal microbial diversity and distinct metagenomic functions. Microbiome 4 (1), 42. Available from: http://microbiomejournal.biomedcentral.com/articles/10.1186/s40168-016-0189-7.

Falony, G., et al., 2016. Population-level analysis of gut microbiome variation. Science 352 (6285), 560–564.

Fonseca, D.M., et al., 2015. Microbiota-dependent sequelae of acute infection compromise tissue-specific immunity. Cell 163 (2), 354–366. Available from: http://www.ncbi.nlm.nih.gov/pubmed/26451485.

Forslund, K., et al., 2015. Disentangling type 2 diabetes and metformin treatment signatures in the human gut microbiota. Nature 528 (7581), 262–266. Available from: http://www.nature.com/doifinder/10.1038/nature15766.

Fujimura, K.E., et al., 2014. House dust exposure mediates gut microbiome Lactobacillus enrichment and airway immune defense against allergens and virus infection. Proc. Natl. Acad. Sci. U. S. A. 111 (2), 805–810. Available from: http://www.ncbi.nlm.nih.gov/pubmed/24344318.

Goodrich, J.K., et al., 2014. Human genetics shape the gut microbiome. Cell 159 (4), 789–799. Available from: http://www.ncbi.nlm.nih.gov/pubmed/25417156.

Goodrich, J.K., et al., 2016. Genetic determinants of the gut microbiome in UK twins. Cell Host Microbe 19 (5), 731–743. Available from: http://www.ncbi.nlm.nih.gov/pubmed/27173935.

Jackson, M.A., et al., 2016. Proton pump inhibitors alter the composition of the gut microbiota. Gut 65 (5), 749–756. Available from: http://www.ncbi.nlm.nih.gov/pubmed/26719299.

Jernberg, C., et al., 2007. Long-term ecological impacts of antibiotic administration on the human intestinal microbiota. ISME J. 1 (1), 56–66. Available from: http://www.nature.com/doifinder/10.1038/ismej.2007.3.

Korpela, K., et al., 2016. Intestinal microbiome is related to lifetime antibiotic use in Finnish pre-school children. Nat. Commun. 7, 10410. Available from: http://www.ncbi.nlm.nih.gov/pubmed/26811868.

Laursen, M.F., et al., 2015. Having older siblings is associated with gut microbiota development during early childhood. BMC Microbiol. 15, 154. Available from: http://www.ncbi.nlm.nih.gov/pubmed/26231752.

Lee, H., Ko, G., 2014. Effect of metformin on metabolic improvement and gut microbiota. Appl. Environ. Microbiol. 80 (19), 5935–5943. Available from: http://www.ncbi.nlm.nih.gov/pubmed/25038099.

Lewis, B.B., et al., 2015. Loss of microbiota-mediated colonization resistance to clostridium difficile infection with oral vancomycin compared with metronidazole. J. Infect. Dis. 212 (10), 1656–1665. Available from: http://www.ncbi.nlm.nih.gov/pubmed/25920320.

Li, J., et al., 2014. An integrated catalog of reference genes in the human gut microbiome. Nat. Biotechnol. 32 (8), 834–841. Available from: http://www.ncbi.nlm.nih.gov/pubmed/24997786.

Lode, H., et al., 2001. Ecological effects of linezolid versus amoxicillin/clavulanic acid on the normal intestinal microflora. Scand. J. Infect. Dis. 33 (12), 899–903. Available from: http://www.ncbi.nlm.nih.gov/pubmed/11868762.

Moeller, A.H., et al., 2016. Cospeciation of gut microbiota with hominids. Science 353 (6297).

Riedler, J., et al., 2001. Exposure to farming in early life and development of asthma and allergy: a cross-sectional survey. Lancet 358 (9288), 1129–1133. Available from: http://www.ncbi.nlm.nih.gov/pubmed/11597666.

Risnes, K.R., et al., 2011. Antibiotic exposure by 6 months and asthma and allergy at 6 years: findings in a cohort of 1,401 US children. Am. J. Epidemiol. 173 (3), 310–318. Available from: https://academic.oup.com/aje/article-lookup/doi/10.1093/aje/kwq400.

Schloss, P.D., et al., 2014. The dynamics of a family's gut microbiota reveal variations on a theme. Microbiome 2 (1), 25. Available from: http://www.microbiomejournal.com/content/2/1/25.

Spanogiannopoulos, P., et al., 2016. The microbial pharmacists within us: a metagenomic view of xenobiotic metabolism. Nat. Rev. Microbiol. 14 (5), 273–287. Available from: http://www.nature.com/doifinder/10.1038/nrmicro.2016.17.

Stein, M.M., et al., 2016. Innate immunity and asthma risk in amish and hutterite farm children. N. Engl. J. Med. 375 (5), 411–421. Available from: http://www.nejm.org/doi/10.1056/NEJMoa1508749.

Strachan, D., et al., 1997. Childhood antecedents of allergic sensitization in young British adults. J. Allergy Clin. Immunol. 99 (1), 6–12. Available from: http://linkinghub.elsevier.com/retrieve/pii/S009167499781038X.

Tong, M., et al., 2014. Reprogramming of gut microbiome energy metabolism by the FUT2 Crohn's disease risk polymorphism. ISME J. 8 (11), 2193–2206. Available from: http://www.ncbi.nlm.nih.gov/pubmed/24781901.

Zhang, J., et al., 2015. A phylo-functional core of gut microbiota in healthy young Chinese cohorts across lifestyles, geography and ethnicities. ISME J. 9 (9), 1979–1990. Available from: http://www.ncbi.nlm.nih.gov/pubmed/25647347.

Zhang, S.L., et al., 2017. Human and rat gut microbiome composition is maintained following sleep restriction. Proc. Natl. Acad. Sci. U. S. A. 114 (8), E1564–E1571. Available from: http://www.ncbi.nlm.nih.gov/pubmed/28179566.

Zhernakova, A., Kurilshikov, A., Bonder, M., Tigchelaar, E., Schirmer, M., et al., 2016. Population-based metagenomics analysis reveals markers for gut microbiome composition and diversity. Science 352 (6285), 565–569.

Zhuang, X., et al., 2017. Alterations of gut microbiota in patients with irritable bowel syndrome: A systematic review and meta-analysis. J. Gastroenterol. Hepatol. 32 (1), 28–38. Available from: http://doi.wiley.com/10.1111/jgh.13471.

CHAPTER 6

Impact of Nutrition on the Gut Microbiota

Objectives

- To learn about the impact of diet on gut microbiota in healthy individuals.
- To understand what is known about the influence of macronutrients (carbohydrate, protein, and fat), dietary patterns, food components, and food additives on the composition and metabolites of the gut microbiota.
- To become acquainted with individual responses to diet that may be driven by gut microbiota.
- To gain perspective on the links between diet, gut microbiota, and health.

Never before has the saying "we are what we eat" been so appropriate. Although health professionals have known for hundreds of years that optimal nutrition leads to health and well-being and that undernutrition is linked to chronic disease, research on the gut microbiota demonstrates that nutrition and health are more closely linked than previously thought. The influence of dietary factors on the gut microbiome is a growing area of interest among nutrition scientists. Not only can nutrition influence the composition of the gut microbiota, but also, the microbiota can change the body's metabolic response to nutrition—with possible ramifications for health and disease.

This chapter focuses on the influence of diet on the gut microbiota of healthy individuals. A growing body of evidence from both observational and experimental studies supports the notion that diet shapes microbial composition. A large cohort study found 63 dietary factors (including those related to carbohydrate, protein, and fat intake) related to differences in gut microbiota composition between individuals (Falony et al., 2016). The insights of the past decade are outlined below; it must be noted that research in this area is at a preliminary stage and the full potential of diet to modulate the microbiome for the maintenance of health is far from clear. At present, the creation of concrete dietary guidelines from these insights is challenging.

Gut Microbiota
https://doi.org/10.1016/B978-0-12-810541-2.00006-3

NUTRITION MODULATES THE MICROBIOTA

Diet is an established modulator of gut microbiota composition, with shifts evident within 24h of a dietary change (David et al., 2013). Various food components, dietary patterns, and nutrients all have the potential to alter the growth of different microbial species and/or change the community dynamics, modulating the microbial population of the colon considerably. While large observational studies show only 16%–19% of the variation in gut microbiota composition from person to person can be explained by intrinsic and environmental factors, diet is one of the most influential environmental aspects (Zhernakova et al., 2016; Falony et al., 2016).

As described in Chapter 2, the majority of human digestion and absorption of nutrients occurs in the upper part of the GI tract. A fraction of normal human dietary intake remains undigested in the small intestine and passes through to the large intestine, where the microbiota break down certain components. The substrate that is fermented by the gut microbiota provides a source of energy for the growth of certain microbes living there, modulating aspects of the microbial community. The host is able to use the resulting metabolites, which may or may not be beneficial to health. The colon is emerging as an important site for investigating many aspects of health, as the complex interactions between host and microbiota—including degradative activities in the colon that yield nutrients and metabolites—are better understood.

Macronutrients Meet the Microbiota

Carbohydrates

Carbohydrates, the major energy production source for the body, are the most well-studied dietary components in relation to the modification of the human gut microbiota. Since "dietary fiber" is an imprecise concept and varying definitions exist around the world, scientists have proposed the term **microbiota-accessible carbohydrates** (MACs) to refer to the types of fiber that can be metabolically used by gut microbes; MACs include "carbohydrates that are dietary and resistant to degradation and absorption by the host, and they may be secreted by the host in the intestine or be produced by microbes within the intestine" (Sonnenburg and Sonnenburg, 2014). MACs may be derived from plants, animals, or food-borne microbes. The intestinal microbiota ferments MACs to produce different metabolically active end products (e.g., short-chain fatty acids or SCFAs). The definition of dietary MACs does not include cellulose and lignin, as these substrates are poorly metabolized by the intestinal microbiota.

Many different types of diet-derived MACs are present in the large intestine at any given time, particularly in the proximal bowel. Concentrations vary as the substrates are utilized, replenished, or replaced through dietary intake. Metabolically versatile bacteria capable of thriving on different carbohydrate sources can adapt to changing nutritional circumstances and flourish in the intestinal ecosystem.

Numerous studies have linked high dietary fiber (MAC) intake with better health: lower body weight, less cardiovascular disease, and improved gastrointestinal health (Slavin, 2013). The mechanisms by which MACs lead to health benefits have yet to be fully elucidated, but a growing body of research shows that gut microbiota may be implicated.

Manipulating overall carbohydrate intake in the diet appears to induce profound changes to the microbiota and to SCFA production. The first trial in humans to demonstrate the relationship between butyrate production and varying carbohydrate content in the diet—carried out with obese individuals—showed that a decreased intake of carbohydrates led to reductions in total SCFA concentrations, with butyrate showing the greatest decrease (Duncan et al., 2007). Significant reductions were also seen in the populations of *Bifidobacterium* spp., *Roseburia* spp., and *Eubacterium rectale* (i.e., notable butyrate producers) when intake of carbohydrates was reduced.

Targeted analyses show diets rich in resistant starch result in an increased abundance of bacteria belonging to the *Clostridium* cluster IV (Ruminococcaceae), whereas diets high in nonstarch polysaccharides result in higher levels of *Clostridium* cluster XIVa (Lachnospiraceae) (Salonen et al., 2014). *Clostridium* spp. are important for colonocytes (colon cells), as they release butyrate as an end product of fermentation. Increased microbial gene richness, a likely marker of health, was seen in healthy obese/overweight subjects that followed an energy-restricted diet with a concomitant increase in soluble fiber consumption (Cotillard et al., 2013). High microbiota diversity and many kinds of complex carbohydrates in the diet are required for the increased production of SCFAs (see Fig. 6.1) and associated health benefits. Chapter 7 gives more information on therapeutic manipulation of the microbiota through nondigestible carbohydrates (such as those that qualify as prebiotics).

Protein

Dietary protein is an important nutrient for health: it is essential for critical processes in growth, immunity, and reproduction. With the assistance of enzymes, mainly proteases and peptidases, dietary protein is primarily

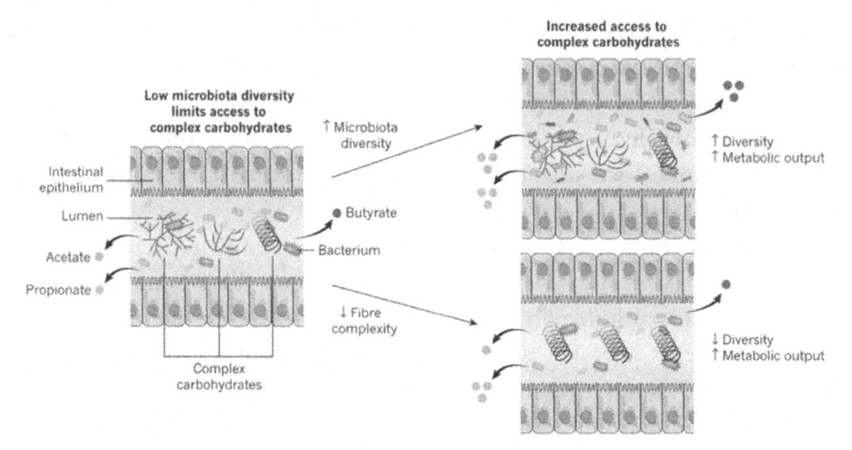

Fig. 6.1 *Top right*: When microbiota diversity is high and many types of complex carbohydrates are consumed in the diet, a higher percentage of these complex carbohydrates are accessible to gut microbes, yielding SCFAs such as acetate, propionate, and butyrate. *Left*: When microbiota diversity is low and the diet contains many complex carbohydrates, only a low percentage is accessible to the microbiota. *Bottom right*: If limited types of complex carbohydrates are supplied in the diet and if this low-fiber diet is matched to the needs of low-diversity microbiota, SCFA production might increase, but microbiota diversity will remain low. *Top right*: Upon the consumption of a complex carbohydrate-rich diet, levels of SCFAs increase and help recruit more diversity to the microbiota (Sonnenburg and Bäckhed, 2016). *(Reproduced with permission from Sonnenburg, J.L., Bäckhed, F., 2016. Diet-microbiota interactions as moderators of human metabolism. Nature 535, 56–64. Macmillan Publishers Ltd: Nature. Copyright 2016.)*

processed into peptides and free amino acids (FAAs) in the small intestine. Enterocytes absorb the peptides and FAAs, and they go on to be metabolized by different organs of the body. In parallel with carbohydrates, the dietary protein, peptides, and FAAs that escape digestion and absorption in the upper GI tract enter the large intestine for further fermentation in the distal colon by the gut microbiota. It is estimated that, in individuals consuming a Western diet, an average of between 6 and 18 g of protein per day reaches the large intestine for microbial fermentation (Cummings and Macfarlane, 1991). Protein that reaches the distal colon is derived from both dietary sources and from the host: dietary sources may include plant protein and animal muscle protein, while host sources of protein in the intestine are in the form of enzymes, mucins, and other glycoproteins from oral, gastric, pancreatic, and small intestinal secretions (Macfarlane and Macfarlane, 2012).

The metabolism of dietary protein by the microbiota is less well defined in the literature than that of carbohydrates. There appear to be two pathways that the microbiota use to metabolize protein sources. First, the bacteria proteolyze (break down) the protein source to FAAs, which are incorporated into structural and other proteins within the microbes (Gibson et al., 1989). Second, the microbiota members ferment the amino acids to generate a large range of metabolites, including hydrogen, methane, carbon dioxide, hydrogen sulfide, SCFAs, and also branched-chain fatty acids (BCFAs), ammonia, *N*-nitroso compounds, amines, and phenolic and indolic compounds (Yao et al., 2016). The metabolites resulting from amino acid fermentation elicit a wide range of biological functions via different receptors and mechanisms; however, animal studies have established that many of these metabolites are detrimental to colonic health (see Fig. 6.2). In vitro studies confirm that ammonia, produced as a protein fermentation metabolite, for example, is harmful to intestinal health (Windey et al., 2012).

Changes to the gut microbiota have also been documented when the supply of dietary protein is increased. *Bacteroides* spp. are highly associated with animal proteins, whereas *Prevotella* spp. are highly associated with increased intakes of plant proteins (Wu et al., 2011). Interventional studies have demonstrated that high-protein diets (animal protein) result in reductions in fecal butyrate concentrations and butyrate-producing bacteria such as *Bifidobacterium* spp., *Roseburia* spp., and *E. rectale* (Brinkworth et al., 2009; Duncan et al., 2007; Russell et al., 2011).

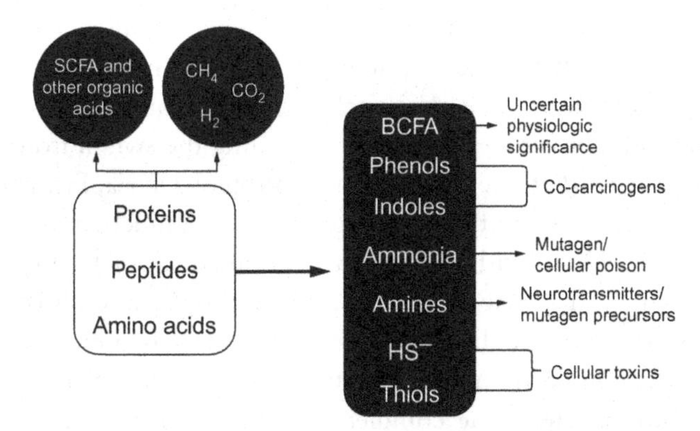

Fig. 6.2 When colonic microbes ferment amino acids, many of the resulting bacterial metabolites have a negative impact on colonic health. *BCFA*, branched-chain fatty acids; *SCFA*, short-chain fatty acids. *(From Macfarlane, G.T., Macfarlane, S., 2012. Bacteria, colonic fermentation, and gastrointestinal health. J. AOAC Int. 95 (1), 50–60.)*

Fecal concentrations of potentially damaging *N*-nitroso compounds were seen to increase markedly in volunteers who consumed a high-protein, low-carbohydrate diet (Russell et al., 2011). Furthermore, a study of five male volunteers consuming high intakes of animal protein demonstrated that fecal sulfide production is related to meat intake (Magee et al., 2000); hydrogen sulfide is a compound associated with ulcerative colitis (Rowan et al., 2009). Overall, it seems that excessive protein intake or an unsuitable ratio between protein and carbohydrate in the diet appears to increase pathogens and protein-fermenting bacteria, with potentially adverse effects on health (Ma et al., 2017).

Fat

Fatty acids help form biological membranes in the body and aid in efficient storage of energy, as well as cellular binding/recognition, signaling, digestion, and metabolism (Janson and Tischler, 2012). In terms of dietary fat intake, most current recommendations around the world suggest avoiding trans fats, limiting saturated fat, and replacing saturated fat with monounsaturated and polyunsaturated fats. The optimal ratio of fatty acid intake is controversial, with the health effects attributable to different sources of dietary fat remaining unclear. The gut microbiota may play a role in some of these health effects, but limited information exists on the impact of fats on the human gut microbiome; the majority of research is limited to animal studies.

Rodent studies have demonstrated that both the quantity and quality of dietary fats impact the microbial ecosystem; while it is not clear at this point whether bacteria use lipids for energy, the presence of lipids in the gut can impact microbial activity. Indeed, dramatic changes in microbial communities occur when lean mice are switched from a standard chow diet to a high-fat diet (Hildebrandt et al., 2009). Following the switch from standard chow to a high-fat diet, the proportions of Proteobacteria, Firmicutes, and Actinobacteria increase, and Bacteroidetes decrease (Hildebrandt et al., 2009; Patterson et al., 2014), with spikes in the populations of Lachnospiraceae (Patterson et al., 2014). The changes are believed to be a result of saturated fatty acid overflow to the distal colon, resulting in diet-induced changes in gut microbiota composition. Another study demonstrated that members of the family Bifidobacteriaceae completely disappear in mice fed a high-fat diet, possibly corroborating other studies (Zhang et al., 2009). Furthermore, Desulfovibrionaceae (associated with more severe obesity) was found to be more prevalent in mice fed a high-fat diet. In mice, it seems that diets high in fat have a negative impact on health through the microbiota and may disrupt gut microbiota in two ways: by diminishing levels of gut barrier-protecting *Bifidobacterium* spp. and by increasing the presence of endotoxins in the blood (Cani et al., 2007).

The type of dietary fatty acid may have more of an influence on the microbiota than total fatty acids. In particular, palm oil (saturated fat) reduces microbial diversity and increases the Firmicutes to Bacteroidetes ratio more significantly than safflower (*n*-6 polyunsaturated fat (PUFA)) or olive oil (monounsaturated fat) in mice (Wit et al., 2012; Patterson et al., 2014). Olive oil and flaxseed/fish oil result in the most diverse intestinal microbiota, with the highest proportions of Bacteroidaceae and *Bacteroides* found in mice supplemented with an olive oil diet (Patterson et al., 2014). A diet rich in flaxseed/fish oil (*n*-3 polyunsaturated fat) appears to have a bifidogenic effect, significantly increasing the intestinal populations of Bifidobacteriaceae and in particular, *Bifidobacterium* spp. The possible mechanism is the increased ability of flaxseed/fish oil to increase the adhesion of bifidobacteria to the intestinal wall (Patterson et al., 2014). Saturated fats (from palm oil) appear to increase total SCFA concentrations the most when compared with diets supplemented with monounsaturated (olive oil) or polyunsaturated fats (safflower or flax/fish oil) (Patterson et al., 2014). SCFAs are generally considered beneficial to health; however, it has been shown in obese individuals that the obesity-associated gut microbiome has increased capacity to harvest energy from these dietary fatty acids (Turnbaugh et al., 2006).

The rise in obesity has triggered an interest in low-carbohydrate, high-fat diets for weight loss in the general public. Intervention trials in humans that have examined the impact of high-fat diets on the intestinal microbiome are limited. Fava et al. (2012) reported that a high-fat diet resulted in a reduction in total bacteria, whereas a diet high in saturated fat increased fecal SCFA concentrations in subjects at risk for metabolic syndrome. A clinical trial comparing a low-carbohydrate, high-fat diet with a high-carbohydrate, high-fiber, low-fat diet in overweight/obese individuals confirmed that a very-low-carbohydrate, high-fat diet resulted in lower fecal concentrations and excretion of SCFAs; also observed was a reduction in counts of *Bifidobacterium* spp. (Brinkworth et al., 2009). Further intervention trials examining the impact of various types of fat on the human intestinal microbiome are needed.

Micronutrients Meet the Microbiota

Micronutrients are vitamins, minerals, and trace elements that are critical to energy metabolism, cellular growth and differentiation, organ function, and immune function. Vitamins such as thiamine, riboflavin, niacin, biotin, pantothenic acid, and folate (B vitamins), as well as vitamin K, can be synthesized by the microbiota (Biesalski, 2016). While vitamins supplied in the diet are absorbed in the small intestine, microbe-produced vitamins are absorbed in the colon.

Little is known about the micronutrients supplied through the diet and their interactions with the microbiota; particularly relevant is how these micronutrients might affect the immune system, intestinal barrier function, and overall health via the gut microbiome. Future research will need to delineate the impact of intestinal micronutrient synthesis and the action of micronutrients on the composition and activity of the microbiota (Biesalski, 2016) in order to uncover the mechanisms of how these are linked to health.

Dietary Pattern Meets the Microbiota

Dietary patterns are defined as "the quantities, proportions, variety or combinations of different foods and beverages in diets, and the frequency with which they are habitually consumed" (Office of Disease Prevention and Health Promotion, 2015). Dietary patterns can vary in nutritional quality and are associated with different health outcomes. In research, dietary patterns can be useful for characterizing the way an individual eats and enables the study of how diet may link to long-term health. While it is known that dietary patterns exert long-term selective pressure on the gut microbiota (Lloyd-Price et al., 2016), a plausible hypothesis, unproven to date, is that the health effects of any dietary pattern are partially attributable to the way it affects gut microbiota composition and/or function.

Western Diet Pattern Compared to Traditional Diet Patterns

The "Western" diet represents a high intake of fat, cholesterol, animal-derived protein, and sugar, with excess salt intake and frequent consumption of processed and convenience foods. It is also characterized by a lower consumption of fruits, vegetables (fibers and micronutrients), and whole cereals (Devereux, 2006; Manzel et al., 2014). This diet is typically consumed in many industrialized countries and has been linked to adverse health outcomes: obesity, metabolic disease, cardiovascular disease, and possibly autoimmune disease (Manzel et al., 2014). The Western diet pattern, for example, is hypothesized to increase pro-inflammatory cytokines, modulate intestinal permeability, and alter the intestinal microbiota in a way that promotes low-grade chronic inflammation in the gut (Huang et al., 2013). In a large-cohort association study, decreased microbiota diversity was correlated with some features of a Western diet, such as higher overall energy intake, snacking, and sugar-sweetened beverage consumption (Falony et al., 2016).

The comparison of dietary patterns between populations in unindustrialized rural communities and industrialized Westernized communities has revealed very specific gut microbiome adaptations to diet and lifestyle. Although multiple lifestyle factors differ between these groups, distinct dietary patterns appear to be responsible for some of the gut microbiome differences.

A foundational study on the impact of a Western diet on the intestinal microbiome was a dietary comparison study between children living in a rural village in Burkina Faso (BF) and children living in an urban area of Florence, Italy (EU) (De Filippo et al., 2010). Children from BF consumed a diet low in fat; low in animal protein; and rich in starch, fiber, and plant polysaccharides. Children in the EU consumed a typical Western diet high in animal protein, sugar, starch, and fat but low in fiber. Common to the microbiota of both groups were representatives of the phyla Actinobacteria, Bacteroidetes, Firmicutes, and Proteobacteria. However, there were profound differences between the two groups in the proportion of bacteria from each phylum. BF children showed a significant enrichment in Bacteroidetes and a reduction in Firmicutes; they had bacteria from the genera *Prevotella* and *Xylanibacter*—known to contain a set of bacterial genes for cellulose and xylan hydrolysis—completely lacking in the EU children. Significantly more fecal SCFAs were found in BF than in EU children. Also, Enterobacteriaceae (*Shigella* and *Escherichia)*, known for causing noninfectious gastrointestinal illness, were significantly underrepresented in BF compared with EU children (De Filippo et al., 2010). Similar results were found when the microbiota of children in Bangladesh was compared with that of children living in an upper-middle-class neighborhood in the United States (Lin et al., 2013). In comparison with US children, the Bangladeshi children had a microbiota that was enriched in *Prevotella*, *Butyrivibrio*, and *Oscillospira* and reduced in *Bacteroides*; these differences were believed to be related to constituents of the diet (e.g., rice, bread, and lentils in the Bangladeshi diet). In these studies and others, bacteria belonging to the genus *Prevotella* stand out as a possible indicators of a non-Western, plant-rich diet.

Another study compared the microbiota of Guahibo Amerindians in Venezuela with those living in rural communities of Malawi and with those in US metropolitan areas (St. Louis, Philadelphia, and Boulder) (Yatsunenko et al., 2012). The US individuals consumed a Western diet, while diets in Malawi and Amerindian populations were dominated by corn and cassava and included some industrial goods, such as soda in Malawi and milk products, canned products, and soda in Venezuela. The individuals living in the three countries showed significant differences in the composition of bacterial species and functional gene repertoires in the gut microbiota, with especially pronounced separation of the US group from the Malawian and Amerindian groups. This finding was evident in early infancy and adulthood. Evidence that these gut microbiome differences were attributable to diet included the observation that genes coding for several enzymes involved in the degradation of amino acids and simple sugars were overrepresented in US adult fecal microbiomes, reflecting their diet pattern high in protein and sugar. In contrast, enzymes participating in the degradation of starch were overrepresented in the Malawian and Amerindian microbiomes, reflecting their corn-rich diet. Another published study supported the results of the above comparisons: residents of rural Papua New Guinea were found to have greater bacterial diversity in their fecal microbiota compared with US residents; this was possibly related to the Papua New Guineans' higher intake of plant-derived carbohydrates and dietary fiber with less frequent intake of meat-derived protein (Martinez et al., 2015). These studies support the notion that, of all the features of a Western diet, it may be the reduced intake of starch and fiber that are most reflected in the gut microbiome.

Hadza individuals from northwestern Tanzania, who practice a foraging lifestyle, consume a diet of wild foods and practice no cultivation or domestication of plants and animals. They were found to have higher levels of microbial richness and biodiversity than individuals from urban Italy who consumed a Mediterranean diet (Schnorr et al., 2014). The Hadza gut microbiota was depleted in *Bifidobacterium*, enriched in Bacteroidetes and *Prevotella*, and characterized by an unusual arrangement of Clostridiale clusters, perhaps reflecting the Hadza people's ability to digest and extract valuable nutrition from fibrous plant foods. Metagenomic analysis demonstrated that the Hadza intestinal microbiome was uniquely adapted for efficient carbohydrate processing and energy capture from plant-derived complex polysaccharides, whereas the Italian gut microbiome was enriched in genes involved in the metabolism of refined carbohydrates (e.g., glucose, galactose, and sucrose) (Rampelli et al., 2015). Then, Obregon-Tito et al. (2015) compared the microbiota of the Matses, a remote hunter-gatherer population from the Peruvian Amazon, with a traditional agricultural community from the Andean highlands called the Tunapuco and with residents of Norman, Oklahoma, a typical US community with an urban-industrialized lifestyle (Obregon-Tito et al., 2015). This study supported previous findings of taxonomic and metabolic differences between those living traditional rural versus Westernized lifestyles. However, the researchers found that the gut microbiota of Matses and Tunapuco individuals were enriched for Spirochaetes, specifically the genus *Treponema*, which was not present in the guts of Americans. *Treponema succinifaciens* is a nonpathogenic carbohydrate metabolizer that is likely to increase when a high-fiber diet, containing tubers, is consumed. The researchers postulated that Spirochaetes may represent a part of the human ancestral gut microbiome that has been lost in Western populations through dietary differences, the adoption of industrial agriculture, and/or other lifestyle changes (Obregon-Tito et al., 2015).

Another illuminating study investigated microbiota composition and function in two Central African Republic populations, BaAka and Bantu, and compared it with those in the United States. BaAka individuals practice a hunter-gatherer lifestyle, with a diet consisting of wild game, vegetables, fruits, and fish. Bantu individuals use agricultural practices, growing tubers and other vegetables, consuming flour-like products, and raising goats; the Bantu thus represent a lifestyle midway between hunter-gatherers and Western individuals. Results showed that the fecal microbiomes of BaAka and Bantu individuals were compositionally similar but the BaAka harbored greater counts of bacteria associated with high intakes of plant fiber, including Prevotellaceae, *Treponema*, and Clostridiaceae; the Bantu showed a domination of Firmicutes. Americans had the smallest numbers of *Prevotella* and *Treponema*. In terms of gut microbiome function, Bantu and Americans both showed an increase in metabolic pathways involved in processing sugar and xenobiotics. Overall, the gut microbiome data in these three populations corresponded with the degree of Westernization in their diets and lifestyles: BaAka and Americans were the most different from each other, with Bantu falling somewhere in between (Gomez et al., 2016).

An interventional study also backed up the observational data on how a Western diet affects the gut microbiome: in a 2-week "dietary switch" study, African-Americans in the United States (Pittsburgh) swapped their habitual diet with rural South Africans. Under close supervision, the African-Americans were supplied with a high-fiber, low-fat (African-style) diet, and the rural Africans were supplied with a high-fat, low-fiber (Western-style) diet. From this short intervention, the researchers found only minor compositional changes in the gut microbiota; however, the Americans that switched to the high-fiber diet showed decreased markers of inflammation, a greater abundance of butyrate synthesizing genes, and higher fecal butyrate, while those on the high-fat diet showed the reverse. Since the changes in these biomarkers are known to affect colorectal cancer (CRC) risk, the researchers pointed out the gut microbiome's possible involvement in the increased CRC risk associated with a Western diet (O'Keefe et al., 2015).

The above studies suggest that individuals living in nonindustrialized, rural societies have a microbiota with significant diversity (within individuals) and lower variation (between individuals), with major compositional differences. These changes may be partially a result of distinct dietary habits. Westernization is associated with a loss of microbial diversity in the gut, including organisms able to ferment fiber-rich dietary components, and increased abundance of bacteria able to ferment animal-based products (Segata, 2015).

Vegetarian Diet Pattern

Vegetarianism is a dietary pattern that is based on the consumption of plants rather than meats. It includes different types of diets that vary on whether they include animal-derived foods such as milk and eggs (do Rosario et al., 2016). Consumption of a vegetarian diet is associated with a number of health benefits such as a significantly lower risk for ischemic heart disease mortality, cancer (Huang et al., 2012), and type 2 diabetes, compared with nonvegetarians (Satija et al., 2016). The health benefits appear to derive from the increased consumption of polyphenols and fibers, combined with restriction of meat and/or animal products; it has recently been hypothesized that this combination creates a specific bacterial niche and leads to the production of distinct metabolites that have diverse abilities to metabolize certain nutrients (do Rosario et al., 2016), leading to robust health.

Two intervention trials have examined the impact of vegetarian diets on the human microbiota (Kim et al., 2013; David et al., 2013). In the first, obese volunteers that followed a strict vegetarian diet for 1 month showed significant changes in the composition of the microbiota compared with baseline (Kim et al., 2013): a significant reduction in the Firmicutes to Bacteroidetes ratio; a reduction in pathobionts; and an enhanced growth of bacteria from the Lachnospiraceae, Ruminococcaceae, and Erysipelotrichaceae families. The subjects also showed reduced body weight, improved markers of metabolic health, and reduced gut inflammation. The second study was a crossover trial where each subject was provided with an animal-based diet (with no fiber) and a plant-based diet (high in fiber), with a washout period between the two diets. The animal-based diet had the greatest impact on the subjects' microbiota, with increased diversity and abundance of bile-tolerant microorganisms (*Alistipes*, *Bilophila*, and *Bacteroides*) and decreased levels of bacteria that metabolize dietary fiber (*Roseburia*, *E. rectale*, and *Ruminococcus bromii*). On the plant-based diet, subjects showed increased populations of *Prevotella* (David et al., 2013). These trials support the notion of increased dietary fiber having positive effects on the gut microbiota and on health.

Several observational studies that examined the gut microbiota of subjects consuming a vegetarian diet versus an omnivorous diet support these diet intervention trials (Liszt et al., 2009; Matijasic et al., 2014; Ruengsomwong et al., 2014; Zimmer et al., 2012; Kabeerdoss et al., 2012; Ferrocino et al., 2015; Reddy et al., 1998). Table 6.1 summarizes findings on the impact of a vegetarian diet on the microbiota as described in observational studies. The different results obtained from these studies may be attributable to methodological differences, but most of the changes observed have been linked with some form of health benefit. Further large-scale interventional studies over longer periods are warranted to confirm these results.

Mediterranean Diet Pattern

The Mediterranean diet pattern (MDP) is rich in plant foods (cereals, fruits, vegetables, legumes, tree nuts, seeds, and olives) and includes moderate intakes of fish and seafood; moderate consumption of eggs, poultry, and dairy products (cheese and yogurt) with low consumption of red meat, processed meat, and sweets; and a moderate intake of alcohol (mainly wine during meals) (Bach-Faig et al., 2011). The principal source of dietary lipids of the MDP is olive oil, a monounsaturated fatty acid. Lifestyle features that often accompany the MDP include daily physical activity and the consumption of seasonally available, local foods (see Fig. 6.3).

Table 6.1 Findings of observational studies on the impact of a vegetarian diet on the microbiota

Common findings in individuals consuming a vegetarian diet
Higher bacterial diversity (Liszt et al., 2009)
Increased levels of *Prevotella* (Matijasic et al., 2014; Ruengsomwong et al., 2014; De Filippo et al., 2010)
Increased levels of *Bacteroides* (Liszt et al., 2009; Matijasic et al., 2014; De Filippo et al., 2010)
Decreased pathobionts, including members of the family Enterobacteriaceae (Kim et al., 2013; De Filippo et al., 2010)
Increased counts of *Faecalibacterium prausnitzii* (Liszt et al., 2009; Matijasic et al., 2014)
Reduced stool pH (Wu et al., 2011; Zimmer et al., 2012)

Modified from do Rosario, V.A., Fernandes, R., de Trindade, E.B.S.M., 2016. Vegetarian diets and gut microbiota: Important shifts in markers of metabolism and cardiovascular disease. Nutr. Rev. 74 (7), 444–454.

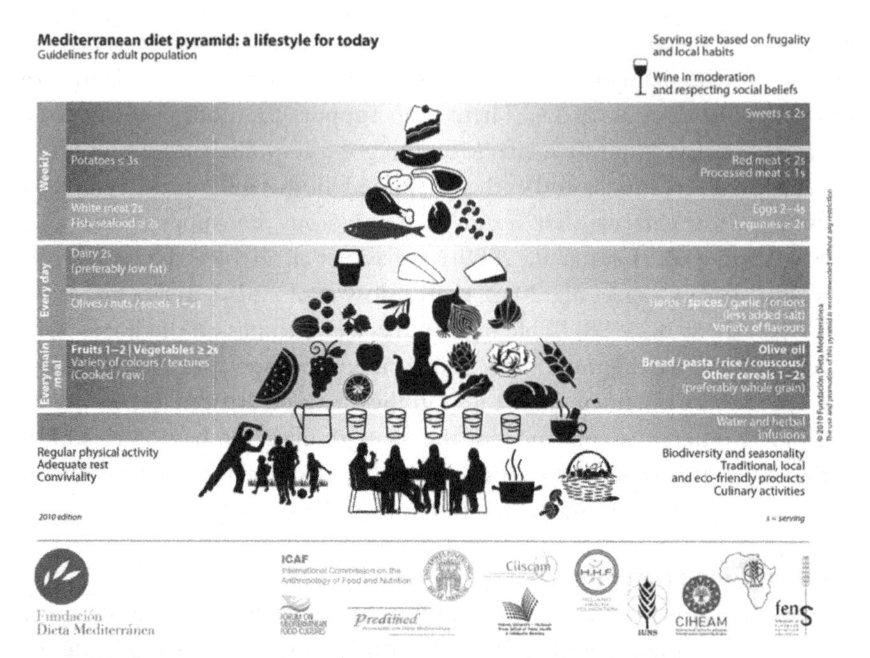

Fig. 6.3 Components of the Mediterranean diet pattern, including its associated lifestyle factors. *(Reproduced with permission from Bach-Faig, A., et al., 2011. Mediterranean diet pyramid today. Science and cultural updates. Public Health Nutr. 14 (12A), 2274–2284.)*

The MDP has been associated with significant improvements in health status, such as a reduced risk of type 2 diabetes (Schwingshackl et al., 2015), cancer (Schwingshackl and Hoffmann, 2014), Alzheimer's disease (Singh et al., 2014), metabolic syndrome (Garcia et al., 2016), and cardiovascular disease (Tong et al., 2016). The MDP is also associated with antiinflammatory properties (Estruch et al., 2006), as improvement in inflammatory markers with consumption of this dietary pattern has been demonstrated in those with disease (Marlow et al., 2013). The protective effect is believed to be derived from the fatty acid profile with emphasis on monounsaturated and polyunsaturated fatty acids and from the polyphenols from increased consumption of plant-based high-fiber foods with low glycemic index (Cuervo et al., 2014).

Accumulating evidence shows the MDP affects the gut microbiome of healthy individuals in a way that may account for its known health benefits (De Filippis et al., 2015; Gutiérrez-Díaz et al., 2016). In one of the first attempts to study the impact of regular adherence to an MDP on the fecal microbiota and its metabolites, an observational study was completed in a sample of healthy adults (Gutiérrez-Díaz et al., 2016). Subjects that had the highest adherence to the MDP had a higher proportion of bacteria in the phylum Bacteroidetes and in the genus *Prevotella* and a lower abundance of those in the phylum Firmicutes and genus *Ruminococcus*. Also, higher concentrations of the fecal SCFAs propionate and butyrate were found in subjects with higher adherence to the MDP. Similar results were discovered in a cross-sectional study of subjects adhering to the MDP (De Filippis et al., 2015): more abundant Bacteroidetes and *Prevotella*. An increase in fecal SCFA levels was also associated with the MDP, which the authors attributed to the higher levels of bacteria belonging to both Firmicutes and Bacteroidetes that metabolize carbohydrates.

The fecal microbiota of patients with metabolic syndrome was analyzed following the consumption of an MDP for 2 years (Haro et al., 2016a). The MDP induced a significant increase in the abundance of *Parabacteroides distasonis*, *Bifidobacterium adolescentis*, and *Bifidobacterium longum* in the subjects with metabolic syndrome, in addition to *Bacteroides thetaiotaomicron* and *Faecalibacterium prausnitzii*, two species with a putative role intestinal homeostasis (Miquel et al., 2013). This indicated that the MDP may increase or maintain a microbiota with antiinflammatory capability. A further study by Haro et al. (2016b) in obese men following an MDP for 1 year showed a diet-linked decrease in *Prevotella*, with increased *Roseburia* and *Oscillospira* genera. The abundance of *P. distasonis* also increased after long-term consumption of the MDP; however, no differences were observed in the main metabolic variables for these men.

The above research indicates that, while adherence to an MDP is associated with possible beneficial effects on the gut microbiome, further studies need to confirm these findings. The MDP, with its high intake of fiber and particular fats, constitutes an intriguing potential approach to enhancing human health by shaping the gut microbiota.

Food Components Meet the Microbiota

Polyphenols

The work described previously in this chapter suggests that plant-based diets are associated with potentially beneficial effects on health and the gut microbiome. The fiber in plants undoubtedly accounts for many of the benefits, but in addition, researchers have identified polyphenols as having potentially unique impacts on health. Dietary polyphenols are natural compounds occurring in plants, including foods such as fruits, vegetables, cereals, tea, coffee, and wine. Flavonoids are a group of polyphenolic compounds that are sometimes studied separately. Around 90%–95% of total polyphenol intake reaches the large intestine, where the colonic microbiota metabolize the polyphenolic structures into low-molecular-weight phenolic metabolites (Cardona et al., 2013). These metabolites are absorbable, and some researchers have posited that they are responsible for the health effects derived from polyphenol-rich food consumption. Evidence from human studies indicates that phenolic compounds in foods can alter gut microbial composition. While a huge array of foods contain potentially health-modifying polyphenols, some of the main sources that have been studied are summarized below and in Table 6.2.

Tea

Green tea has long been recognized in East Asia for its beneficial impact on health; benefits may be derived from flavonoids, which occur in tea in large quantities. The major classes of flavonoids found in tea are catechins, including epicatechin, epigallocatechin, epicatechin-3-gallate, and epigallocatechin-3-gallate (Etxeberria et al., 2013). A product containing 70% tea polyphenols (equivalent to 10 cups of concentrated green tea) resulted in a gut microbiota with a significant reduction in *Clostridium perfringens* and other *Clostridium* spp. with a significant increase in *Bifidobacterium* spp. (Okubo et al., 1992). SCFA concentrations, specifically propionate and acetate, also increased significantly with the tea polyphenol intake. In agreement with this study, fecal samples of 10 volunteers who drank green tea instead of water for 10 days showed an increase in the proportion of *Bifidobacterium* spp., with a concomitant decrease after green tea ingestion was discontinued (Jin et al., 2012). An in vitro study that exposed human fecal homogenates to various flavonoids in green tea found that certain pathogenic bacteria such as *C. perfringens*, *Clostridium difficile*, and *Bacteroides* spp. were significantly repressed by tea phenolics and their derivatives while commensal anaerobes like *Clostridium* spp., *Bifidobacterium* spp., and *Lactobacillus* spp. were less severely affected (Lee et al., 2006). Different strains of intestinal bacteria had varying degrees of growth sensitivity to the various tea phenolics and metabolites. Few human studies have examined the impact of black tea on the gut microbiome.

Red Wine

Consuming moderate amounts of red wine has been shown to have beneficial health effects, and a large body of work has attributed these to its phenolic compounds. Red wine consists not only of a complex mixture of flavonoids, such as flavan-3-ols (referred to as flavanols) and anthocyanins, but also of nonflavonoids such as resveratrol, cinnamates, and gallic acid (Etxeberria et al., 2013). Large-scale population data showed that higher

Table 6.2 A summary of the human studies that demonstrate the impact of polyphenols on gut microbiota

Phenolic compound	Amount	Impact on microbiome	Reference
Green tea (flavanols)	~10 cups/day	Decrease in *Clostridium perfringens* and other *Clostridium* spp., increase in *Bifidobacterium* spp., increase in SCFAs	Okubo et al. (1992)
Green tea (flavanols)	4 cups/day	Increase in *Bifidobacterium* spp.	Jin et al. (2012)
Red wine: dealcoholized red wine and red wine	272 mL/day	Increased *Enterococcus, Prevotella, Bacteroides, Bifidobacterium, Bacteroides uniformis, Eggerthella lenta*, and *Blautia coccoides-E. rectale* group	Queipo-Ortuño et al. (2012)
Cocoa	494 mg/day	Increase in *Eubacterium rectale-C. coccoides* group, *Lactobacillus* spp., *Enterococcus* spp., *Bifidobacterium* spp.	Tzounis et al. (2011)
Wild blueberry	25 g/day	Increase in *Lactobacillus acidophilus* and *Bifidobacterium* spp.	Vendrame et al. (2011)
Soy	100 mg/day of isoflavones aglycon equivalents	Increase in Erec cluster, *Lactobacillus-Enterococcus, Faecalibacterium prausnitzii*, and *Bifidobacterium* spp.	Clavel et al. (2005)
Soy milk (26.5% beta-conglycinin/ 38.7% glycinin)	500 mL/day	Decreased Firmicutes to Bacteroidetes ratio, increased *Eubacterium* and *Clostridium*, decreased *Bifidobacterium* spp.	Fernandez-Raudales et al. (2012)

consumption of red wine leads to greater abundance of the antiinflammatory species *F. prausnitzii* (Falony et al., 2016). Also, the microbiota of 10 healthy volunteers was compared after consecutive intake of red wine, dealcoholized red wine, and gin for 4 weeks (Queipo-Ortuño et al., 2012). After the period of consuming red wine, there was a significant increase in *Enterococcus, Prevotella, Bacteroides, Bifidobacterium*, and several additional groups in stool samples. Within Firmicutes, the genus *Enterococcus* and the *Blautia coccoides-E. rectale* group increased significantly after consumption of dealcoholized red wine and red wine compared with baseline. Within Bacteroidetes, the genus *Bacteroides* and the *B. uniformis* species and the number of *Prevotella* increased significantly after red wine intake. Within the Actinobacteria phylum, red wine and dealcoholized red wine led to a significant increase in the number of *Bifidobacterium* and *Eggerthella lenta* compared with baseline. This study showed that red wine consumption can significantly modulate the growth of potentially beneficial bacteria in the gut microbiota of humans, which suggests a possible role of the gut microbiome in the health benefits associated with the inclusion of red wine in the diet (Queipo-Ortuño et al., 2012).

Cocoa

Cocoa, a product derived from *Theobroma cacao* L. (Sterculiaceae), is rich in flavanol compounds (flavan-3-ols). Very little is known about the effects of these compounds on the human gut microbiota, although Tzounis et al. (2011) observed a significant difference in the population amounts of *Bifidobacterium* spp., the *Clostridium histolyticum* group, *E. rectale-C. coccoides* group, and *Lactobacillus* and *Enterococcus* spp. following the consumption of a cocoa-containing drink (494 mg/day) (Tzounis et al., 2011).

Fruit

Berries contain abundant phenolic compounds, mostly flavonoids (where anthocyanins predominate) (Etxeberria et al., 2013). Following 6 weeks of consumption of a wild blueberry (*Vaccinium angustifolium*) drink, healthy volunteers showed significant increases in *Bifidobacterium* spp. and *Lactobacillus acidophilus* (species with possible health benefits) (Vendrame et al., 2011).

Soy

Products derived from soybeans (members of Leguminosae) are rich in phytoestrogens, principally in the form of isoflavones (Etxeberria et al., 2013). This class of flavonoids has been shown to alter gut bacterial composition and diversity in postmenopausal women (Clavel et al., 2005): after 1 month of supplementation with isoflavones (100 mg/day isoflavones aglycone equivalents), Erec cluster, *Lactobacillus-Enterococcus, F. prausnitzii* subgroup, and the genus *Bifidobacterium* significantly increased (Clavel et al., 2005). Organisms that are part of the Erec cluster are known to metabolize isoflavones.

Soybean protein isolate, which is primarily composed of glycinin and beta-conglycinin proteins (50%–70% of total protein), has been shown to have beneficial effects on health (Xiao, 2008). Supplementation of low-glycinin soy milk (49.5% beta-conglycinin/6% glycinin) and conventional soy milk (26.5% beta-conglycinin/38.7% glycinin) in overweight and obese men led to changes in microbial composition (Fernandez-Raudales et al., 2012). The relative abundance of Firmicutes significantly decreased, whereas the relative abundance of Bacteroidetes significantly increased following consumption of both low-glycinin and conventional soy milk. The genus *Faecalibacterium* was more abundant in the low-glycinin soy milk group, whereas the genera *Eubacterium* and *Clostridium* were more abundant in the conventional soy milk group. Contrary to other studies, *Bifidobacterium* was significantly reduced following the consumption of both low-glycinin and conventional soy milk (Fernandez-Raudales et al., 2012).

This research confirms that dietary polyphenols appear to exert prebiotic-like effects that contribute to the maintenance of health through the gut microbiota. Although there are plenty of in vitro studies, data on the impact of polyphenols on the gut microbiota and mechanisms of action in humans are lacking. Further studies are required to create an enhanced understanding of the relationship between dietary phenolics and gut microbiota, with a combination of metagenomics and metabolomics studies, to provide more insight into the health effects of polyphenols.

Food Additives Meet the Microbiota

Emulsifiers

Dietary emulsifiers contribute to the desirable characteristics of many processed foods and beverages while remaining indigestible, unabsorbable, and unfermentable (Glade and Meguid, 2016). Carboxymethylcellulose and polysorbate-80 are used in various foods at concentrations up to 2% (Cani and Everard, 2015). There is emerging evidence to suggest that emulsifiers are associated with alterations in gut microbiota composition and can increase bacterial translocation, possibly promoting diseases associated with gut inflammation, such as inflammatory bowel disease and metabolic syndrome (Chassaing et al., 2015). Rodents predisposed to colitis that were provided with low concentrations of carboxymethylcellulose and polysorbate-80 (ranging from 0.1% to 1% of food per 24 h) exhibited low-grade inflammation, obesity/metabolic syndrome, and increased colitis (Chassaing et al., 2015). The emulsifiers weakened the mucus barrier of the intestinal epithelium and facilitated bacterial translocation into the intestinal tissues. In addition, the ingestion of low-dose emulsifiers promoted subtle signs of chronic intestinal inflammation, including epithelial damage. Extrapolating these results to humans is challenging as the mice received continual doses of the emulsifiers, which resulted in levels of these compounds that are probably an overestimation of what humans would consume (Cani, 2015). Further studies emulating realistic human levels and the mechanisms and effects on obesity, low-grade inflammation, and microbiota are necessary.

Noncaloric Sweeteners

Noncaloric sweeteners (NCS), also known as artificial sweeteners or high-intensity sweeteners, are food additives used to replace sugar in food and impart a sweet taste. In the United States and Canada, these include acesulfame potassium, aspartame, saccharin, steviol glycosides, monk fruit extract, sucralose, and neotame. Although NCS provide a sweet taste, they deliver few or no calories and are consumed by millions worldwide as means of combating weight gain and maintaining glycemic control; paradoxically, however, they have been associated with weight gain (Pepino and Bourne, 2011). The majority of NCS are not absorbed in the body; they are excreted and considered metabolically inert (Roberts et al., 2000). Despite the fact that NCS are not metabolized by human digestive mechanisms, emerging evidence from animal models suggests that NCS are metabolized by the intestinal microbiota. Therefore, these compounds may impact the gut microbiome and host health (Suez et al., 2015).

Suez and colleagues supplemented the drinking water of mice with saccharin, sucralose, or aspartame and found that each of the mice that consumed NCS displayed marked glucose intolerance as compared with controls, with saccharin having the most pronounced effect. Mice drinking saccharin had a distinct microbiome characterized by the enrichment of taxa belonging to the *Bacteroides* genus or the Clostridiales order, with an underrepresentation of lactobacilli and other members of the Clostridiales. The microbiota appeared to be responsible for the observed effects, since only the transfer of feces from NCS-consuming mice into germ-free mice induced impaired glucose tolerance in recipients. A human component of the study compared high NCS consumers with non-NCS consumers and found positive correlations between multiple taxonomic groups in the gut and NCS consumption (expansion of Actinobacteria, the Enterobacteriales order, and the Clostridiales order with increased NCS intake). NCS consumption was positively correlated with various clinical parameters such as higher body mass index and increased blood pressure, hemoglobin A1c, and fasting glucose levels. Also, in seven healthy volunteers who did not normally consume NCS, after 1 week of saccharin supplementation (upper daily intake of 5 mg/kg/day), four out of seven developed poorer glycemic responses (i.e., the "responders"), and the rest showed no significant changes. The responders showed compositional changes in their microbiota. Stool from these subjects was transferred into mice both before and after the NCS consumption; only the stool from after NCS consumption induced glucose intolerance in the recipient mice. The authors hypothesized that individuals feature a personalized response to NCS, possibly stemming from initial differences in microbiota composition and function.

Research examining NCS and the gut microbiome is in its early phases. It is not known how NCS besides saccharin and aspartame impact the gut microbiome. Further, human clinical trials are needed to determine the full extent to which NCS may affect the human microbiome and health.

INDIVIDUALIZED RESPONSES TO DIET

Several recent studies have shown apparent individualized responses to diet. In one study, healthy individuals showed different glucose responses to a high-fiber barley-based bread, with *Prevotella/Bacteroides* being associated with a beneficial response. When the gut microbiota of those with the most positive response was transferred to germ-free mice, the rodents showed improved glucose metabolism and increased *Prevotella* abundance (Kovatcheva-Datchary et al., 2015).

In an important study from Israel, researchers monitored glucose levels and food consumption in more than 800 people over 1 week and found surprisingly high variability in glucose response to the same food from person to person. Some individuals showed a spike in blood glucose after consuming foods such as tomatoes, not typically considered high-glycemic-index foods. The researchers then developed a machine-learning algorithm based on microbiome data and other features, which allowed accurate prediction of glucose response to a given food; short-term personalized dietary interventions that they developed were able to normalize postmeal glucose responses in a smaller cohort (Zeevi et al., 2015). These studies show that the gut microbiota may be critical to an individual's physiological response to food and that studies of gut microbiota modulation by diet need to account for baseline microbiota composition and/or function.

DIETARY CHANGES IN POPULATIONS OVER TIME

An intriguing line of research is showing that the gut microbiome adapts to make use of new dietary components when eating habits change over time. For example, Japanese individuals have a gut bacterium called *Bacteroides plebeius* with genes coding for carbohydrate-active enzymes that help the body gain energy from a form of seaweed (Hehemann et al., 2010). These bacterial genes are absent from North American individuals. It appears that frequent dietary intake of seaweeds with their associated marine bacteria in Japanese individuals succeeded in adding new functions to the gut microbiome that aid in seaweed digestion (Hehemann et al., 2012). Thus, dietary change over time may contribute to the gut microbial genes that help extract energy from the components.

FUTURE DIRECTIONS

It is clear from both observational and intervention studies that food components and dietary patterns shape the gut microbiome. Consumption of diets rich in plant-based foods versus diets with a high intake of animal fat and protein result in notable differences in microbial composition. A diet rich in plant-based foods, including fiber and polyphenols, appears to be beneficial to gut health because of the substrates (vitamins, SCFAs, and other products) provided to the host through microbial fermentation. Although a highly diverse microbiota seems a worthy goal for dietary interventions, multispecies conversions of dietary components must be taken into account in future studies to fully understand the impact of diet on health. Moreover, dietary intervention studies will help strengthen causal connections between dietary aspects and health-related microbial modulation (Zoetendal and de Vos, 2014).

REFERENCES

Bach-Faig, A., et al., 2011. Mediterranean diet pyramid today. Science and cultural updates. Public Health Nutr. 14 (12A), 2274–2284.

Biesalski, H.K., 2016. Nutrition meets the microbiome: micronutrients and the microbiota. Ann. N.Y. Acad. Sci. 1372 (1), 53–64.

Brinkworth, G.D., et al., 2009. Comparative effects of very low-carbohydrate, high-fat and high-carbohydrate, low-fat weight-loss diets on bowel habit and faecal short-chain fatty acids and bacterial populations. Br. J. Nutr. 101 (10), 1493–1502.

Cani, P.D., 2015. Metabolism: dietary emulsifiers—sweepers of the gut lining? Nat. Rev. Endocrinol. 11 (6), 319–320.

Cani, P.D., Everard, A., 2015. Keeping gut lining at bay: impact of emulsifiers. Trends Endocrinol. Metab. 26 (6), 273–274.

Cani, P.D., et al., 2007. Selective increases of bifidobacteria in gut microflora improve high-fat-diet-induced diabetes in mice through a mechanism associated with endotoxaemia. Diabetologia 50 (11), 2374–2383.
Cardona, F., et al., 2013. Benefits of polyphenols on gut microbiota and implications in human health. J. Nutr. Biochem. 24 (8), 1415–1422.
Chassaing, B., et al., 2015. Dietary emulsifiers impact the mouse gut microbiota promoting colitis and metabolic syndrome. Nature 519 (7541), 92–96.
Clavel, T., et al., 2005. Isoflavones and functional foods alter the dominant intestinal microbiota in postmenopausal women. J. Nutr. 135 (12), 2786–2792.
Cotillard, A., et al., 2013. Dietary intervention impact on gut microbial gene richness. Nature 500 (7464), 585–588.
Cuervo, A., et al., 2014. Pilot study of diet and microbiota: interactive associations of fibers and polyphenols with human intestinal bacteria. J. Agric. Food Chem. 62 (23), 5330–5336.
Cummings, J.H., Macfarlane, G.T., 1991. The control and consequences of bacterial fermentation in the human colon. J. Appl. Bacteriol. 70 (6), 443–459.
David, L.A., et al., 2013. Diet rapidly and reproducibly alters the human gut microbiome. Nature 505 (7484), 559–563. Available from: http://www.nature.com/doifinder/10.1038/nature12820.
De Filippis, F., et al., 2015. High-level adherence to a Mediterranean diet beneficially impacts the gut microbiota and associated metabolome. Gut. gutjnl-2015-309957.
De Filippo, C., et al., 2010. Impact of diet in shaping gut microbiota revealed by a comparative study in children from Europe and rural Africa. Proc. Natl. Acad. Sci. U. S. A. 107 (33), 14691–14696.
Devereux, G., 2006. The increase in the prevalence of asthma and allergy: food for thought. Nat. Rev. Immunol. 6, 869–874.
do Rosario, V.A., Fernandes, R., de Trindade, E.B.S.M., 2016. Vegetarian diets and gut microbiota: important shifts in markers of metabolism and cardiovascular disease. Nutr. Rev. 74 (7), 444–454.
Duncan, S.H., et al., 2007. Reduced dietary intake of carbohydrates by obese subjects results in decreased concentrations of butyrate and butyrate-producing bacteria in feces. Appl. Environ. Microbiol. 73 (4), 1073–1078.
Estruch, R., et al., 2006. Effects of a Mediterranean-style diet on cardiovascular risk factors a randomized trial. Ann. Intern. Med. 145 (1), 1–11.
Etxeberria, U., et al., 2013. Impact of polyphenols and polyphenol-rich dietary sources on gut microbiota composition. J. Agric. Food Chem. 61 (40), 9517–9533.
Falony, G., et al., 2016. Population-level analysis of gut microbiome variation. Science 352 (6285), 560–564.
Fava, F., et al., 2012. The type and quantity of dietary fat and carbohydrate alter faecal microbiome and short-chain fatty acid excretion in a metabolic syndrome "at-risk" population syndrome. Int. J. Obes. 37 (2), 216–223.
Fernandez-Raudales, D., et al., 2012. Consumption of different soymilk formulations differentially affects the gut microbiomes of overweight and obese men. Gut Microbes 3 (6), 490–500.
Ferrocino, I., et al., 2015. Fecal microbiota in healthy subjects following omnivore, vegetarian and vegan diets: culturable populations and rRNA DGGE profiling. PLoS One 10 (6), e0128669.
Garcia, M., et al., 2016. The effect of the traditional Mediterranean-style diet on metabolic risk factors: a meta-analysis. Nutrients 8 (168), 1–18.
Gibson, S.A.W., et al., 1989. Significance of microflora in proteolysis in the colon. Appl. Environ. Microbiol. 55 (3), 679–683.
Glade, M.J., Meguid, M.M., 2016. A glance at … dietary emulsifiers, the human intestinal mucus and microbiome, and dietary fiber. Nutrition 32 (5), 609–614.

Gomez, A., et al., 2016. Gut microbiome of coexisting BaAka pygmies and bantu reflects gradients of traditional subsistence patterns. Cell Rep. 14 (9), 2142–2153. Available from: http://linkinghub.elsevier.com/retrieve/pii/S2211124716300997.

Gutiérrez-Díaz, I., et al., 2016. Mediterranean diet and faecal microbiota: a transversal study. Food Funct. 2347–2356.

Haro, C., Garcia-Carpintero, S., et al., 2016a. The gut microbial community in metabolic syndrome patients is modified by diet. J. Nutr. Biochem. 27, 27–31.

Haro, C., Montes-Borrego, M., et al., 2016b. Two healthy diets modulate gut microbial community improving insulin sensitivity in a human obese population. J. Clin. Endocrinol. Metab. 101 (1), 233–242.

Hehemann, J.-H., et al., 2010. Transfer of carbohydrate-active enzymes from marine bacteria to Japanese gut microbiota. Nature 464 (7290), 908–912. Available from: http://www.nature.com/doifinder/10.1038/nature08937.

Hehemann, J.-H., et al., 2012. Bacteria of the human gut microbiome catabolize red seaweed glycans with carbohydrate-active enzyme updates from extrinsic microbes. Proc. Natl. Acad. Sci. U. S. A. 109 (48), 19786–19791. Available from: http://www.ncbi.nlm.nih.gov/pubmed/23150581.

Hildebrandt, M.A., et al., 2009. High-fat diet determines the composition of the murine gut microbiome independently of obesity. Gastroenterology 137 (5), 1716–1724.

Huang, T., et al., 2012. Cardiovascular disease mortality and cancer incidence in vegetarians: a meta-analysis and systematic review. Ann. Nutr. Metab. 60 (4), 233–240.

Huang, E.Y., et al., 2013. The role of diet in triggering human inflammatory disorders in the modern age. Microbes Infect. 15 (12), 765–774.

Janson, L., Tischler, M., 2012. Medical Biochemistry: The Big Picture. McGraw-Hill Education, Columbus, OH.

Jin, J.-S., et al., 2012. Effects of green tea consumption on human fecal microbiota with special reference to Bifidobacterium species. Microbiol. Immunol. 56 (11), 729–739.

Kabeerdoss, J., et al., 2012. Faecal microbiota composition in vegetarians: comparison with omnivores in a cohort of young women in southern India. Br. J. Nutr. 108 (6), 953–957.

Kim, M.-S., et al., 2013. Strict vegetarian diet improves the risk factors associated with metabolic diseases by modulating gut microbiota and reducing intestinal inflammation. Environ. Microbiol. Rep. 5 (5), 765–775.

Kovatcheva-Datchary, P., et al., 2015. Dietary fiber-induced improvement in glucose metabolism is associated with increased abundance of Prevotella. Cell Metab. 22 (6), 971–982. Available from: http://linkinghub.elsevier.com/retrieve/pii/S1550413115005173.

Lee, H.C., et al., 2006. Effect of tea phenolics and their aromatic fecal bacterial metabolites on intestinal microbiota. Res. Microbiol. 157 (9), 876–884.

Lin, A., et al., 2013. Distinct distal gut microbiome diversity and composition in healthy children from Bangladesh and the United States. PLoS One 8 (1), e53838.

Liszt, K., et al., 2009. Characterization of bacteria, clostridia and Bacteroides in faeces of vegetarians using qPCR and PCR-DGGE fingerprinting. Ann. Nutr. Metab. 54 (4), 253–257.

Lloyd-Price, J., Abu-Ali, G., Huttenhower, C., 2016. The healthy human microbiome. Genome Med. 8 (1), 51. Available from: http://genomemedicine.biomedcentral.com/articles/10.1186/s13073-016-0307-y.

Ma, N., et al., 2017. Contributions of the interaction between dietary protein and gut microbiota to intestinal health. Curr. Protein Pept. Sci. 18 (999), 1. Available from: http://www.eurekaselect.com/openurl/content.php?genre=article&doi=10.2174/1389203718666170216153505.

Macfarlane, G.T., Macfarlane, S., 2012. Bacteria, colonic fermentation, and gastrointestinal health. J. AOAC Int. 95 (1), 50–60.
Magee, E.A., et al., 2000. Contribution of dietary protein to sulfide production in the large intestine: an in vitro and a controlled feeding study in humans. Am. J. Clin. Nutr. 72 (6), 1488–1494.
Manzel, A., et al., 2014. Role of "Western diet" in inflammatory autoimmune diseases. Curr. Allergy Asthma Rep. 14 (1), 404. Available from: http://www.ncbi.nlm.nih.gov/pubmed/24338487.
Marlow, G., et al., 2013. Transcriptomics to study the effect of a Mediterranean-inspired diet on inflammation in Crohn's disease patients. Hum. Genomics 7, 24.
Martinez, I., et al., 2015. The gut microbiota of rural papua new guineans: composition, diversity patterns, and ecological processes. Cell Rep. 11 (4), 527–538.
Matijasic, B.B., et al., 2014. Association of dietary type with fecal microbiota in vegetarians and omnivores in Slovenia. Eur. J. Nutr. 53 (4), 1051–1064.
Miquel, S., et al., 2013. *Faecalibacterium prausnitzii* and human intestinal health. Curr. Opin. Microbiol. 16 (3), 255–261.
Obregon-Tito, A.J., et al., 2015. Subsistence strategies in traditional societies distinguish gut microbiomes. Nat. Commun. 6, 6505.
Office of Disease Prevention and Health Promotion, 2015. Dietary patterns—2015 Advisory Report—health.gov. Scientific report of the 2015 Dietary Guidelines Advisory Committee. Available from: https://health.gov/dietaryguidelines/2015-scientific-report/07-chapter-2/.
O'Keefe, S.J.D., et al., 2015. Fat, fibre and cancer risk in African Americans and rural Africans. Nat. Commun. 6, 6342. Available from: http://www.nature.com/doifinder/10.1038/ncomms7342.
Okubo, T., et al., 1992. In vivo effects of tea polyphenol intake on human intestinal microflora and metabolism. Biosci. Biotechnol. Biochem. 56 (4), 588–591.
Patterson, E., et al., 2014. Impact of dietary fatty acids on metabolic activity and host intestinal microbiota composition in C57BL/6J mice. Br. J. Nutr. 111 (11), 1905–1917.
Pepino, M.Y., Bourne, C., 2011. Non-nutritive sweeteners, energy balance, and glucose homeostasis. Curr. Opin. Clin. Nutr. Metab. Care 14 (4), 391–395. Available from: http://www.ncbi.nlm.nih.gov/pubmed/21505330.
Queipo-Ortuño, M.I., et al., 2012. Influence of red wine polyphenols and ethanol on the gut microbiota ecology and biochemical biomarkers. Am. J. Clin. Nutr. 95 (6), 1323–1334.
Rampelli, S., et al., 2015. Metagenome sequencing of the Hadza hunter-gatherer gut microbiota. Curr. Biol. 25 (13), 1682–1693.
Reddy, S., et al., 1998. Faecal pH, bile acid and sterol concentrations in premenopausal Indian and white vegetarians compared with white omnivores. Br. J. Nutr. 79 (6), 495–500.
Roberts, A., et al., 2000. Sucralose metabolism and pharmacokinetics in man. Food Chem. Toxicol. 38 (Suppl. 2), 31–41.
Rowan, F.E., et al., 2009. Sulphate-reducing bacteria and hydrogen sulphide in the aetiology of ulcerative colitis. Br. J. Surg. 96 (2), 151–158.
Ruengsomwong, S., et al., 2014. Senior Thai fecal microbiota comparison between vegetarians and non-vegetarians using PCR-DGGE and real-time PCR. J. Microbiol. Biotechnol. 24 (8), 1026–1033.
Russell, W.R., et al., 2011. High-protein, reduced-carbohydrate weight-loss diets promote metabolite profiles likely to be detrimental to colonic health. Am. J. Clin. Nutr. 93 (5), 1062–1072.
Salonen, A., et al., 2014. Impact of diet and individual variation on intestinal microbiota composition and fermentation products in obese men. ISME J. 8 (11), 2218–2230.
Satija, A., et al., 2016. Plant-based dietary patterns and incidence of type 2 diabetes in US men and women: results from three prospective cohort studies. PLoS Med. 13 (6), e1002039.

Schnorr, S.L., et al., 2014. Gut microbiome of the Hadza hunter-gatherers. Nat. Commun. 5, 3654.
Schwingshackl, L., Hoffmann, G., 2014. Adherence to Mediterranean diet and risk of cancer: a systematic review and meta-analysis of observational studies. Int. J. Cancer 135 (8), 1884–1897.
Schwingshackl, L., et al., 2015. Adherence to a Mediterranean diet and risk of diabetes: a systematic review and meta-analysis. Public Health Nutr. 18 (7), 1292–1299.
Segata, N., 2015. Gut microbiome: westernization and the disappearance of intestinal diversity. Curr. Biol. 25 (14), R611–R613.
Singh, B., et al., 2014. Association of Mediterranean diet with mild cognitive impairment and Alzheimer's disease: a systematic review and meta-analysis. J. Alzheimers Dis. 39 (2), 271–282.
Slavin, J., 2013. Fiber and prebiotics: mechanisms and health benefits. Nutrients 5 (4), 1417–1435. Available from: http://www.ncbi.nlm.nih.gov/pubmed/23609775.
Sonnenburg, J.L., Bäckhed, F., 2016. Diet-microbiota interactions as moderators of human metabolism. Nature 535, 56–64.
Sonnenburg, E.D., Sonnenburg, J.L., 2014. Starving our microbial self: the deleterious consequences of a diet deficient in microbiota-accessible carbohydrates. Cell Metab. 20 (5), 779–786.
Suez, J., et al., 2015. Non-caloric artificial sweeteners and the microbiome: findings and challenges. Gut Microbes 6 (2), 149–155.
Tong, T.Y.N., et al., 2016. Prospective association of the Mediterranean diet with cardiovascular disease incidence and mortality and its population impact in a non-Mediterranean population: the EPIC-Norfolk study. BMC Med. 14 (1), 135.
Turnbaugh, P.J., et al., 2006. An obesity-associated gut microbiome with increased capacity for energy harvest. Nature 444 (7122), 1027–1131. Available from: http://www.nature.com/doifinder/10.1038/nature05414.
Tzounis, X., et al., 2011. Prebiotic evaluation of cocoa-derived flavanols in healthy humans by using a randomized, controlled, double-blind, crossover intervention study. Am. J. Clin. Nutr. 93 (1), 62–72.
Vendrame, S., et al., 2011. Six-week consumption of a wild blueberry powder drink increases Bifidobacteria in the human gut. J. Agric. Food Chem. 59 (24), 12815–12820.
Windey, K., de Preter, V., Verbeke, K., 2012. Relevance of protein fermentation to gut health. Mol. Nutr. Food Res. 56 (1), 184–196.
Wit, N.D., et al., 2012. Saturated fat stimulates obesity and hepatic steatosis and affects gut microbiota composition by an enhanced overflow of dietary fat to the distal intestine. Am. J. Physiol. Gastrointest. Liver Physiol. 303 (5), G589–G599.
Wu, G.D., et al., 2011. Linking long-term dietary patterns with gut microbial enterotypes. Science 334 (6052), 105–108.
Xiao, C.W., 2008. Health effects of soy protein and isoflavones in humans. J. Nutr. 138 (6), 1244S–1249S.
Yao, C.K., Muir, J.G., Gibson, P.R., 2016. Review article: insights into colonic protein fermentation, its modulation and potential health implications. Aliment. Pharmacol. Ther. 43 (2), 181–196.
Yatsunenko, T., et al., 2012. Human gut microbiome viewed across age and geography. Nature 486 (7402), 222–227.
Zeevi, D., et al., 2015. Personalized nutrition by prediction of glycemic responses. Cell 163 (5), 1079–1094. Available from: http://www.ncbi.nlm.nih.gov/pubmed/26590418.
Zhang, C., et al., 2009. Interactions between gut microbiota, host genetics and diet relevant to development of metabolic syndromes in mice. ISME J. 4, 232–241.

Zhernakova, A., et al., 2016. Population-based metagenomics analysis reveals markers for gut microbiome composition and diversity. Science 352 (6285), 565–569.
Zimmer, J., et al., 2012. A vegan or vegetarian diet substantially alters the human colonic faecal microbiota. Eur. J. Clin. Nutr. 66 (1), 53–60.
Zoetendal, E.G., de Vos, W.M., 2014. Effect of diet on the intestinal microbiota and its activity. Curr. Opin. Gastroenterol. 30 (2), 189–195. Available from: http://www.ncbi.nlm.nih.gov/pubmed/24457346.

CHAPTER 7

Therapeutic Manipulation of Gut Microbiota

Objectives

- To gain an overview of the therapeutic methods for manipulating the gut microbiome, which include probiotics, prebiotics, fecal microbiota transplantation and microbial consortia, and microbiota-modulating drugs.
- To understand the current definitions of probiotics and prebiotics and the evidence that exists for their therapeutic use.
- To become aware of how probiotics may be employed preventatively in healthy individuals.

In healthy individuals, diet is a modulator of gut microbiota composition with possible effects on health (as discussed in Chapter 6). This chapter gets more specific about gut microbiota modulation, covering the known tools for manipulating the gut microbiome to impart specific improvements in health status in the context of disease. Probiotics, prebiotics, fecal microbiota transplantation or defined microbial consortia, microbiota-modulating drugs, as well as several other interventions are discussed below. The details below demonstrate that manipulation of the microbiota through various means holds significant promise for medical management of disease. Further research is required to advance this very important field.

The interventions listed in this chapter are indeed linked with changes in human health status; however, the mechanisms involving the gut microbiota are not always clear. The efficacy of fecal microbiota transplantation in recurrent *Clostridium difficile* infection, for instance, is likely (but not certainly) achieved through a remodeling of gut microbiota composition. Many more well-designed studies are required to determine whether the observed health outcomes discussed below are directly attributable to the changes effected on the gut microbiome.

Another challenge in this field is that the intestinal microbiota at baseline is not often taken into account, despite recent studies suggesting it may have an impact on the efficacy of a gut microbiota-modulating intervention.

Gut Microbiota
https://doi.org/10.1016/B978-0-12-810541-2.00007-5

As one of many examples, the low-FODMAP diet is an intervention that both modulates the gut microbiota (Halmos et al., 2015) and reduces symptoms in about half of those with IBS (Eswaran et al., 2016); recent data showed that the low-FODMAP diet responders in a group of children with IBS could be predicted by measuring baseline gut microbiota (Chumpitazi et al., 2015).

PROBIOTICS

The well-known definition of **probiotics** was published in a Food and Agriculture Organization of the United Nations and World Health Organization (FAO/WHO) joint report in 2001 (FAO/WHO, 2001), with an update by expert consensus in 2014 (Hill et al., 2014): "live microorganisms that, when administered in adequate amounts, confer a health benefit on the host." A key point is that the microorganisms must have a scientifically demonstrated beneficial effect on health.

Many live microorganisms do not qualify as probiotics. Excluded from the definition are the live microorganisms in traditional fermented foods, since they are uncharacterized (see Fig. 7.1). Furthermore, in the case of fermented foods, scientists have difficulty separating the health benefits conferred by the food matrix from those conferred by the live microbes themselves (Hill et al., 2014). The microbial consortia from human stool that are used in fecal microbiota transplantation are also not included in the definition of probiotics; these mixtures include unknown taxa (bacteria, yeasts, parasites, and viruses), and scientists struggle to identify which of these microbes are responsible for any beneficial health effects (Hill et al., 2014).

Probiotics have an excellent safety profile, but the risk of adverse events is not zero. Marteau (2001) outlined four classes of possible side effects of probiotic use: systemic infections, detrimental metabolic effects, cytokine-mediated immunologic adverse events in susceptible individuals, and transfer of antibiotic resistance genes. Although these are not normally a concern for healthy individuals, particular caution should be exercised with probiotic use in immunocompromised individuals.

Scientists have difficulty elucidating the preventative effects of probiotics (see "Probiotics in Health" box in this chapter) because clear end points may be elusive. The benefits of probiotics are clearer when it comes to disease: improvement in a clinical parameter for a group receiving a probiotic compared with those receiving a placebo is evidence of efficacy. Yet, many

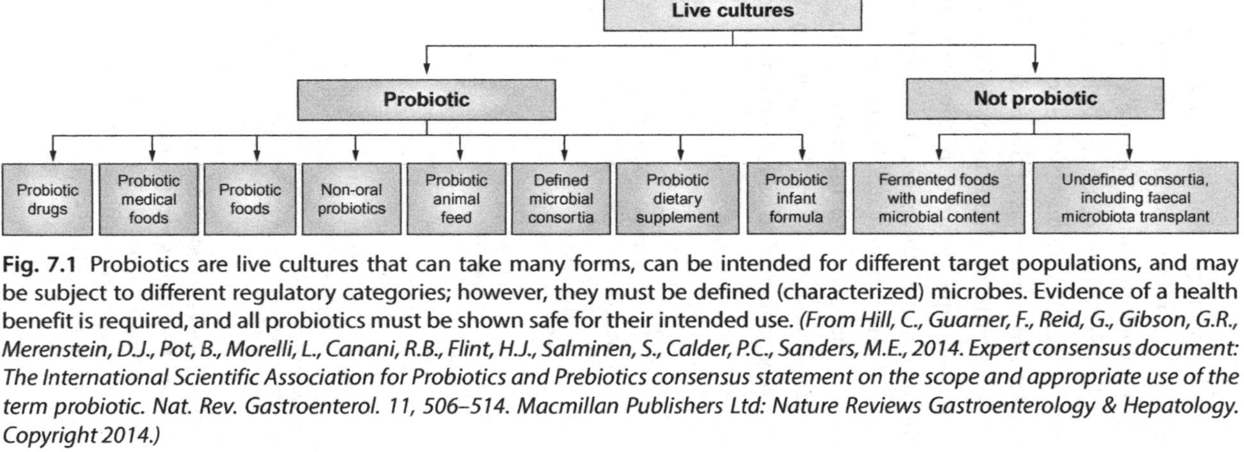

Fig. 7.1 Probiotics are live cultures that can take many forms, can be intended for different target populations, and may be subject to different regulatory categories; however, they must be defined (characterized) microbes. Evidence of a health benefit is required, and all probiotics must be shown safe for their intended use. *(From Hill, C., Guarner, F., Reid, G., Gibson, G.R., Merenstein, D.J., Pot, B., Morelli, L., Canani, R.B., Flint, H.J., Salminen, S., Calder, P.C., Sanders, M.E., 2014. Expert consensus document: The International Scientific Association for Probiotics and Prebiotics consensus statement on the scope and appropriate use of the term probiotic. Nat. Rev. Gastroenterol. 11, 506–514. Macmillan Publishers Ltd: Nature Reviews Gastroenterology & Hepatology. Copyright 2014.)*

metaanalyses have noted that the majority of published studies on probiotics have methodological issues that make it difficult for clinicians to gain reliable insights from the data.

An additional challenge in probiotic research is that a mechanism of action involving the gut microbiota is not confirmed in the vast majority of cases. Clinical studies track probiotic "inputs" (whether a single strain or multiple strains) and health "outputs," often without knowing what happens in between. While scientists do understand some probiotic activities in the gastrointestinal tract and other areas of the body (see Fig. 7.2), very little is generally known about the mechanisms by which probiotics produce their health effects. For example, contrary to the common assumption that probiotic species should colonize the gastrointestinal tract when consumed, a systematic review by Kristensen et al. (2016) found no evidence that probiotic supplementation changed fecal microbiota composition compared with placebo. Thus, simply colonizing and "crowding out" pathogenic bacteria is likely not the mechanism by which probiotics exert their effects.

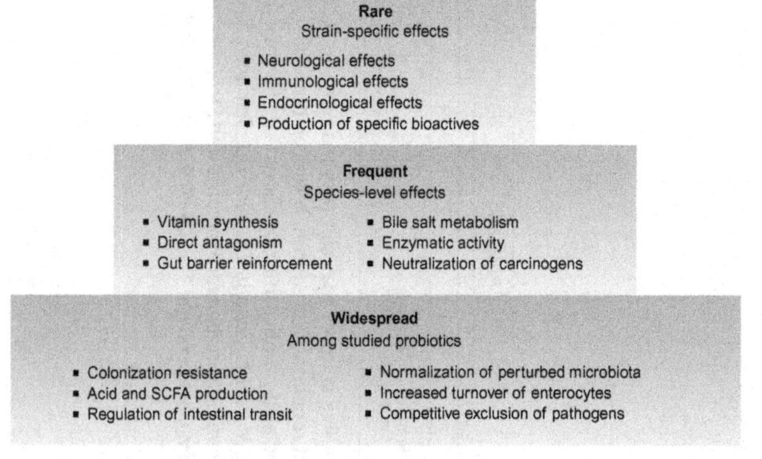

Fig. 7.2 Some probiotic activities are widespread among the probiotic genera that are commonly studied (bottom tier); others are frequently observed in most strains of a species (middle tier); others may occur in only a few strains of a species (top tier). Much more evidence, however, is required to link these activities to the observed health benefits of probiotics. Short-chain fatty acid (SCFA). *(From Hill, C., Guarner, F., Reid, G., Gibson, G.R., Merenstein, D.J., Pot, B., Morelli, L., Canani, R.B., Flint, H.J., Salminen, S., Calder, P.C., Sanders, M.E., 2014. Expert consensus document: The International Scientific Association for Probiotics and Prebiotics consensus statement on the scope and appropriate use of the term probiotic. Nat. Rev. Gastroenterol. 11, 506–514. Macmillan Publishers Ltd: Nature Reviews Gastroenterology & Hepatology. Copyright 2014.)*

Gastrointestinal Function

Evidence suggests probiotics may be beneficial for treating and/or preventing a range of gastrointestinal conditions, which are detailed below. Probiotic strain is a key factor to consider when employing probiotics for a certain indication (Ritchie et al., 2012). The high-quality evidence that exists for probiotic effects on various gastrointestinal diseases is summarized hereafter.

Diarrhea and Constipation

A Cochrane systematic review found that probiotics were a safe and effective way to prevent *C. difficile*-associated diarrhea, reducing the risk by around 64%; however, probiotics were not necessarily effective in reducing the incidence of *C. difficile* infection (Goldenberg et al., 2013).

A review of randomized, controlled trials in adults and children showed that probiotics were effective for treating acute infectious diarrhea: those who took probiotics showed a reduced risk of diarrhea at 3 days and a decreased mean duration of diarrhea by about 30 h. The authors concluded that probiotics could be a useful adjunct to standard rehydration therapy for infectious diarrhea (e.g., rotavirus infection) (Allen et al., 2003). In children, data also support the idea that probiotics might reduce the duration of acute diarrhea and hospital stays in children with acute gastroenteritis (Dinleyici and Vandenplas, 2014).

In studies of children with persistent diarrhea (e.g., those from developing countries), evidence exists for probiotic efficacy, but the data fail to support a dramatic therapeutic effect. That is, probiotics may reduce the duration of diarrhea by around 4 days, with a possible reduction in stool frequency and no adverse events (Bernaola Aponte et al., 2010).

In both adults and children, probiotics may aid in alleviating constipation. Some efficacy is observed for probiotics in increasing bowel movement frequency in adults with functional constipation (Ojetti et al., 2014); a randomized, controlled trial in infants with chronic constipation found that probiotic administration positively affected bowel movement frequency but not stool consistency (Coccorullo et al., 2010).

Inflammatory Bowel Disease

Preliminary evidence shows a possible benefit of probiotics for those with ulcerative colitis (UC). A metaanalysis found probiotics increase remission rates in those with active UC, with no adverse events (Shen et al., 2014). Supplementation with a multispecies probiotic appears safe as an adjunct to standard pharmaceutical treatment in mild-to-moderate UC (Tursi et al.,

2010). Several controlled trials have investigated probiotics for the maintenance of remission in UC and found that they may be as effective as the antiinflammatory drug mesalamine (but no more effective than the drug) for maintaining remission (Verna and Lucak, 2010).

On the other hand, metaanalyses (e.g., Rahimi et al., 2008) show no evidence that probiotics are effective for maintaining remission or for preventing recurrence of Crohn's disease. Probiotics should not be used in this population without further investigation. The different efficacy of probiotics in Crohn's disease versus UC deserves more study and may support the different pathophysiology of the two conditions.

Pouchitis is another condition that may improve with probiotic treatment. Metaanalyses show the benefit of probiotics for managing this condition, with greater efficacy for certain probiotic formulations (Elahi et al., 2008; Nikfar et al., 2010).

Irritable Bowel Syndrome

A metaanalysis of 21 randomized, controlled trials found that probiotics (compared with placebo) in those with irritable bowel syndrome (IBS) were associated with a greater improvement when it came to overall symptom response and quality of life; however, individual symptoms were not reliably affected by probiotics. Interestingly, the improvements in IBS were associated with single probiotic species, lower doses, and short durations of treatment. More evidence is required to elucidate the utility of probiotics for those with IBS (Zhang et al., 2016b).

Cardiovascular Risk Factors

Probiotics are associated with clinical improvements in several factors related to cardiovascular risk. A metaanalysis of 30 randomized, controlled trials concluded that those treated with probiotics showed reduced total cholesterol (by 7.8 mg/dL) and low-density lipoprotein (LDL) cholesterol (7.3 mg/dL) compared with controls, with no differences in high-density lipoprotein (HDL) cholesterol or triglycerides. Probiotics were more effective in individuals with higher baseline total cholesterol levels and when they were administered for longer durations; also, certain probiotic strains were associated with more benefit (Cho and Kim, 2015). Another metaanalysis found probiotic consumption appeared to reduce the cardiovascular-related factors of total cholesterol, LDL, body mass index (BMI), waist circumference, and inflammatory markers, with a notable reduction in LDL with trials of *Lactobacillus acidophilus* (Sun and Buys, 2015). Existing studies do not yet support specific strain and dose recommendations in these populations.

When it comes to blood pressure, a systematic review of nine trials found probiotic consumption changed systolic blood pressure by −3.56 mmHg and diastolic blood pressure by −2.38 mmHg. A greater reduction was observed with multiple probiotic species instead of a single species. Thus, probiotics may improve blood pressure by a modest degree, especially when baseline blood pressure is elevated and when they are taken for at least 8 weeks at an adequate dose (Khalesi et al., 2014).

Metabolic Parameters

Probiotics are probably not an effective strategy for weight reduction: one metaanalysis showed no significant effect of probiotics on body weight or BMI, although authors acknowledged that the methodological quality of the studies was too low to make definitive conclusions (Park and Bae, 2015).

In those with type 2 diabetes, probiotics may be a strategy for improving a number of metabolic parameters: a metaanalysis of eight trials found a significant effect of probiotics on reducing hemoglobin A1c levels and HOMA insulin resistance (a quantification of insulin resistance and beta-cell function) but no effects on other parameters, including fasting plasma glucose (Kasińska and Drzewoski, 2015). Another metaanalysis found probiotics did decrease fasting blood glucose and hemoglobin A1c in those with type 2 diabetes, but this depended on characteristics of each person (e.g., BMI) and the particular probiotic strain and dose (Akbari and Hendijani, 2016).

Depression

Major depressive disorder (or depression) is a psychological state characterized by a low mood that impairs functioning in daily life. In the first systematic review and metaanalysis on probiotics for depression, probiotics were associated with a reduction in depressive symptoms as measured on various rating scales, especially for those aged 60 or younger (Huang et al., 2016); the authors noted that probiotics may also help reduce the risk of depression in nondepressed individuals but that more study is required to confirm this effect. Another systematic review found support for some probiotics in reducing both depression and anxiety in humans, but the authors underlined the need to investigate mechanisms in order to zero in on effective therapeutics (Pirbaglou et al., 2016).

Probiotic use for depression represents only one of several applications under investigation for modulating aspects of brain function. A **psychobiotic** was originally proposed as a subcategory of probiotics; that is, "a live organism that, when ingested in adequate amounts, produces a health benefit in patients suffering from psychiatric illness" (Dinan et al., 2013). Lately, however, some scientists have argued for expansion of the category of psychobiotics, to include both probiotics and prebiotics and other means of influencing the gut microbiome to produce positive mental health effects (Sarkar et al., 2016). To date, the efficacy of probiotics in addressing aspects of brain function and behavior has mainly been shown in animal models.

Infant Health

Regurgitation

Limited evidence indicates that probiotics may have a benefit for regurgitation in infants. One study, for instance, found that administration of *Lactobacillus reuteri* DSM 17938 was beneficial for preventing episodes of regurgitation in breastfed infants in their first month of life (Garofoli et al., 2014).

Infantile Colic

Probiotics have been suggested as a safe strategy for management of infantile colic. Two recent systematic reviews found that infants supplemented with probiotics (*L. reuteri*) showed reductions in mean crying time (Harb et al., 2016; Schreck Bird et al., 2016).

Necrotizing Enterocolitis

A Cochrane systematic review showed that probiotic supplementation significantly reduced the incidence of severe necrotizing enterocolitis (stage II or more) and mortality, without increasing infections; however, efficacy was not confirmed in extremely low-birth-weight infants. The authors say that evidence supports a change in practice for the care of premature infants, although formulation and dose are still matters of debate (AlFaleh et al., 2011).

NEXT-GENERATION PROBIOTICS

In the future, bacterial strains singled out from those in the human gut microbial community may be tested and administered as drugs—these are the so-called **next-generation probiotics**, designer probiotics, or "bugs as drugs."

Scientists have already identified several bacteria mined from the human gut that show promise as next-generation probiotics. Perhaps the most advanced contender is *Akkermansia muciniphila*, a mucin degrading bacterium residing in the intestinal mucus layer that is markedly reduced in those with obesity and type 2 diabetes (Cani and Van Hul, 2015). After many studies of the bacterium in animal models, in 2016, the first human safety trial of live *A. muciniphila* (grown on a synthetic medium) was reported, with therapeutic potential for targeting human obesity and associated metabolic problems (Plovier et al., 2016).

Faecalibacterium prausnitzii is another promising candidate for a next-generation probiotic. As a highly metabolically active, butyrate-producing commensal from the Firmicutes phylum that is abundant in the guts of healthy humans, *F. prausnitzii* may be an indicator of intestinal health. Low levels are reliably observed in those with inflammatory bowel disease (IBD). Mouse studies have shown *F. prausnitzii* administration has antiinflammatory effects and can protect against a form of colitis, and although mechanisms remain unclear, these bacteria show potential for addressing IBD and possibly mixed-type irritable bowel syndrome (Miquel et al., 2013).

As scientists develop these next-generation probiotic therapies, they may need to go beyond administration of the live microorganisms and ensure conditions that enable survival of these bacteria in the human gastrointestinal tract. Furthermore, as these products evolve, they will pose a challenge for regulators, since live microorganisms as drugs may be subject to different standards from conventional drugs.

PREBIOTICS

Prebiotics are another group of compounds used as a tool for microbiota modulation. They are centered around the idea of providing growth substrates for gut microorganisms. At present, compounds that qualify as prebiotics overlap with those in the category of dietary fiber—but importantly, not all dietary fibers are prebiotics because not all of them bring about specific changes in the gut microbiota.

A definition of prebiotics was first advanced by Gibson and Roberfroid in 1995: "nondigestible food ingredients that beneficially affect the host by selectively stimulating the growth and/or activity of one or a limited number of bacterial species already resident in the colon, and thus attempt to improve host health" (Gibson and Roberfroid, 1995). Several updates to the concept have ensued in the years since, with most proposed definitions specifying that prebiotics must target health-promoting groups of bacteria (primarily from the genera *Bifidobacterium* and *Lactobacillus*) or beneficial metabolic activities. A panel of experts discussed an updated definition at a 2010 meeting (Gibson et al., 2010), but the requirement of specific effects on health-promoting microbes broke down because of the difficulty scientists have in categorically identifying beneficial and detrimental members of gut microbiota, and furthermore, microbial community diversity is associated with health more often than the abundance of a particular species.

To overcome these shortcomings in the previous definition(s), a long-awaited consensus definition for prebiotics was published in 2017: "a substrate that is selectively utilized by host microorganisms[,] conferring a health benefit" (Gibson et al., 2017) (see Fig. 7.3). This definition requires that a candidate prebiotic compound acts as a substrate for gut microorganisms and that the beneficial physiological effect depends on the compound's use by microbes. The revision shifts focus away from individual species like bifidobacteria and lactobacilli, but the range of affected microorganisms in a particular host must be limited in order to meet the selectivity criterion (i.e., the measured change must be more specific than a completely altered ecosystem). Indeed, Scott et al. (2013) noted that bacterial cross-feeding

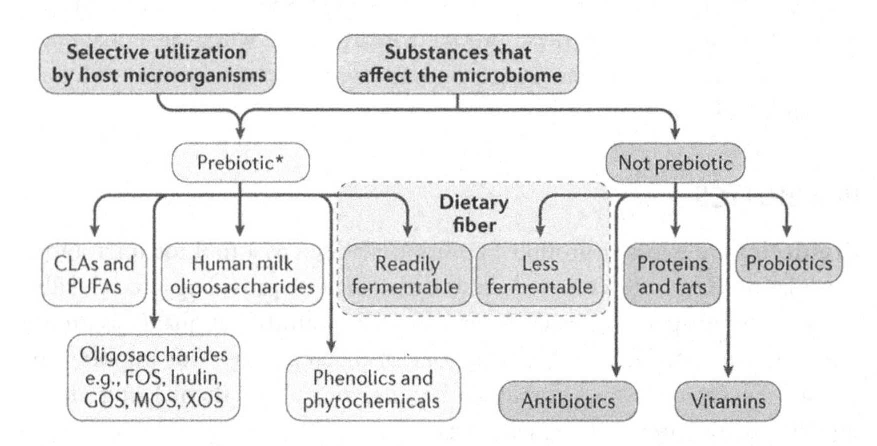

Fig. 7.3 Distinguishing prebiotics from other substances under the new consensus definition. Prebiotics must be selectively utilized by host microorganisms and have evidence of health benefit for the target host (whether human or animal). Dietary prebiotics must not be degraded by the target host's enzymes. The figure shows candidate as well as accepted prebiotics. *CLA*, conjugated linoleic acid; *FOS*, fructooligosaccharides; *GOS*, galactooligosaccharides; *MOS*, mannanoligosaccharide; *PUFA*, polyunsaturated fatty acid; *XOS*, xylooligosaccharide. *(From Gibson, G.R., Hutkins, R., Sanders, M.E., Prescott, S.L., Reimer, R.A., Salminen, S.J., Scott, K., Stanton, C., Swanson, K.S., Cani, P.D., Verbeke, L., Reid, G., 2017. Expert consensus document: The International Scientific Association for Probiotics and Prebiotics (ISAPP) consensus statement on the definition and scope of prebiotics. Nat. Rev. Gastroenterol. Hepatol. Macmillan Publishers Ltd: Nature Reviews Gastroenterology & Hepatology. Copyright 2017.)*

(one species living off the products of another species) is a possible contributor to the prebiotic effect and that the bacteria enriched by prebiotics extend beyond those initially identified.

All current prebiotics are carbohydrates, but it is possible that other compounds could qualify. Under the new definition, future prebiotics need not be restricted to carbohydrates nor even to dietary compounds. In addition, the concept of prebiotics could possibly be applied to body sites outside the gastrointestinal tract (Gibson et al., 2017).

With the consensus definition of prebiotics now established, a contrast with probiotics should be noted. While the definition of probiotic (outlined above) encompasses two concepts—a substance and a health effect—the definition of prebiotic encompasses three: a substance, a health (physiological) effect, and a mechanism.

Traditionally, important prebiotics have included inulin, **fructooligosaccharides** (FOS), and **galactooligosaccharides** (GOS), with several other possible candidates that require more study. **Inulin** is a polysaccharide that naturally occurs in many plants and is often industrially extracted from chicory. FOS, sometimes called oligofructose, are oligosaccharides found in food sources or produced commercially by inulin degradation; they are often used as low-intensity sweeteners or to replace fat content in foods. Inulin and FOS are fructans that are naturally found in certain readily available foods like onion, celery, asparagus, Jerusalem artichokes, and chicory roots, but when administered in specific quantities, they may have therapeutic effects as described below. Also in the prebiotic category are GOS: a mixture of substances produced from a lactose substrate, comprising between two and eight saccharide units. GOS are most commonly used in infant nutrition applications (Torres et al., 2010). Another possible prebiotic candidate is **lactulose**, a synthetically produced sugar consisting of galactose and fructose units, which can be metabolized by microbes. Many studies have observed the capacity of each of these prebiotic compounds to influence gut microbiota in purportedly beneficial ways, but studies on the prebiotics' concomitant effects on host health are few.

Prebiotics are considered very safe; for example, a safety evaluation of FOS by the Nordic Working Group on Food Toxicology and Risk Evaluation concluded that adverse effects like flatulence, abdominal pain, bloating, and diarrhea may occur with human FOS consumption, but for the average person, these effects are unlikely to occur with consumption of 20 g FOS/day or less (Nordic Council of Ministers, 2000).

Gastrointestinal Function

Prebiotics may have immunomodulatory potential in the gut through their production of SCFAs, but clinical studies that have documented the link between prebiotics and specific health outcomes related to immune function are rare. Most promise appears to exist in the area of gastrointestinal health.

Inflammatory Bowel Disease

One review identified the potential of prebiotics for the treatment of inflammatory bowel disease (Macfarlane et al., 2006), although more trials are warranted and specific recommendations are not possible at this time.

Lactose Intolerance

A recent clinical trial found highly purified short-chain GOS led to an increase in the relative abundance of lactose-fermenting *Bifidobacterium*, *Faecalibacterium*, and *Lactobacillus* in the stool of those with lactose intolerance; these changes correlated with improved lactose tolerance (Azcarate-Peril et al., 2017).

Irritable Bowel Syndrome

Previous studies have suggested a possible benefit of prebiotics for IBS; in one recent trial of those with IBS ($n=44$), GOS not only stimulated gut bifidobacteria but also changed stool consistency and improved flatulence, bloating, and overall symptom score; at a higher dose, it also improved anxiety scores (Silk et al., 2009).

Constipation

At both extremes of the lifespan, prebiotics may be effective for alleviating constipation. Healthy infants who received GOS-supplemented formula in their first 12 months showed a softer consistency of stool and increased frequency of defecation, which accompanied changes in the gut microbiota composition (Sierra et al., 2015). Similar results were found in a study of infant supplementation of GOS and FOS combined (Costalos et al., 2008). In elderly individuals with constipation, GOS appeared to increase defecation frequency but responses differed from person to person (Teuri and Korpela, 1998).

Calcium Absorption in Iron Deficiency

Animal studies (e.g., Ohta et al., 1995) show that FOS feeding increases the absorption of calcium, magnesium, and iron under conditions of iron deficiency; there is also preliminary evidence in healthy humans suggesting that inulin may increase calcium absorption (Coudray et al., 1997).

SYNBIOTICS

Synbiotics—combinations of at least one probiotic and one prebiotic ingredient—are so named because of the putative synergistic effects of the components. In theory, synbiotics could benefit the host by providing both probiotics and their preferred growth substrates, enhancing probiotic survival in the gastrointestinal tract. Although not a matter of consensus at present, some argue that the term "should be reserved for products in which the prebiotic compound selectively favors the probiotic compound" (Schrezenmeir and de Vrese, 2001).

Ascertaining the combined effects of any two nutritional ingredients is scientifically complex; little data exist on the use of synbiotics compared with the use of probiotics and prebiotics separately. Moreover, while good adherence of a probiotic on intestinal epithelial cells or intestinal mucus is deemed beneficial, one study found commercially available prebiotics tended to decrease the adherence of probiotic strains to different types of substrate (Kadlec et al., 2014).

Metabolic Parameters

Some synbiotics may have a beneficial effect on metabolic parameters and obesity, but evidence is inconclusive. A systematic review of trials found possible immunomodulatory action exerted by some synbiotics in overweight/obese individuals, with applications to the treatment of metabolic endotoxemia requiring further study (Chiu et al., 2015).

ANTIBIOTICS

The discussion of antibiotics in Chapter 5 focused on cases where they have inadvertent effects on the gut microbiome when taken for a specific indication. But recently, it has emerged that antibiotics and other drugs can also be used to purposefully modulate the microbiome in order to produce a desired health outcome. Antibiotics have a well-known role as therapeutic agents for infectious disease, but they also have newly discovered therapeutic benefits in some noncommunicable diseases (Ianiro et al., 2016). These are outlined below; however, it must be emphasized that given the growing awareness of possible long-term risks of antibiotic use and of antibiotic resistance, clinicians may not always consider modulation of the gut microbiome with antibiotics as the best course of action for treating disease.

Drug-related factors—antibiotic class, dosage, duration of exposure, and route of administration—may be relevant to antibiotic-induced gut microbiota alterations (Ianiro et al., 2016) and how they impact disease status. The research to date is outlined below.

Inflammatory Bowel Disease

Although in current practice antibiotics are recommended only in the case of infections or complications in IBD, metaanalyses show the benefits of antibiotics over placebo for inducing remission. For example, a metaanalysis from 2011 found various kinds of antibiotics (alone or in combination) were superior to placebo for inducing remission in active Crohn's disease, with antibiotics also inducing remission in active UC (Khan et al., 2011). A metaanalysis from 2006 suggested that broad-spectrum antibiotics in particular improved clinical outcomes in individuals with Crohn's disease, although further trials are required (Rahimi et al., 2006). Scientists hypothesize that antibiotics could ameliorate IBDs by decreasing bacterial concentrations or specific bacterial groups in the lumen or by reducing bacterial translocation (Sartor et al., 2004).

Irritable Bowel Syndrome

Rifaximin, an oral, nonsystemic, broad-spectrum antibiotic targeting the gut, shows some efficacy in treating IBS. In two randomized, controlled trials in patients with IBS (without constipation), a 2-week course of rifaximin gave individuals adequate relief of global IBS symptoms (abdominal pain, bloating, and loose or watery stools) for up to 10 weeks (Pimentel et al., 2011).

Hepatic Encephalopathy

Antibiotics are a standard tool for the management of hepatic encephalopathy (HE). Rifaximin, given with or without lactulose, showed higher efficacy than lactulose alone for the treatment of patients with overt HE (Sharma et al., 2013) and was also more effective than placebo in maintaining remission from HE (Bass et al., 2010).

OTHER DRUGS

Evidence is emerging that some drugs owe their therapeutic benefits, at least in part, to the way they modulate the gut microbiota. While at present the list of these drugs is short, this intriguing line of study could lead to an increased number of drugs that target gut microorganisms, rather than the human host, for addressing disease—that is, "drugs that target bugs."

Cyclophosphamide

Emerging evidence shows that gut microbiota helps shape the effects of cyclophosphamide, an "old-school" drug used to treat several types of cancers by stimulating the antitumor immune response. A mouse study by Viaud et al. (2013) showed that cyclophosphamide alters small intestinal gut microbiota composition, leading to the generation of immune cell subsets needed for its antitumor efficacy. This was backed by further work showing that, in mice, the drug's anticancer effect was enhanced by the two commensal species *Enterococcus hirae* and *Barnesiella intestinihominis*, and in humans, immune responses specific to *E. hirae* and *B. intestinihominis* predicted longer progression-free survival in those with lung and ovarian cancer who were being treated with chemoimmunotherapy (Daillère et al., 2016).

New Immunotherapy Drugs

Recently, cancer care has been revolutionized by immunotherapy drugs that are potent stimulators of antitumor T-cell responses. It is becoming clear that specific members of the gut microbiota influence the efficacy of these drugs. In a study of one of these compounds, cytotoxic T lymphocyte antigen-4 (CTLA-4) blockade, researchers found in both mice and humans that T-cell responses specific for *Bacteroides thetaiotaomicron* or *Bacteroides fragilis* were associated with greater drug efficacy. The drug did not work in germ-free mice, showing a key role for specific *Bacteroides* spp. in the drug's immunostimulatory effects (Vetizou et al., 2015).

Metformin

Metformin is a commonly prescribed therapeutic agent for those with type 2 diabetes. The drug's ability to modulate the gut microbiota—in particular, to increase bacteria of the genus *Akkermansia*—could contribute to its ability to improve insulin response (Shin et al., 2014). Forslund et al. (2015) studied a large cohort of people with type 2 diabetes and found that the gut microbiome (specifically, its functional potential for producing butyrate and propionate) mediated the therapeutic effect of the drug, as metformin increased butyrate-producing taxa in the gut microbiota.

FECAL MICROBIOTA TRANSPLANTATION

Fecal microbiota transplantation (FMT) is the transfer of a fecal preparation from a healthy donor to another individual. Although the mechanism of action is still unknown, in theory, FMT works by supplying a stable—though uncharacterized—microbial community to repopulate the colon. Anecdotal reports exist of the application of FMT to a great number of conditions, but scientifically, only a few show promise at present (as described below). Self-administration of FMT may be possible, but health professionals strongly advise against it since the long-term risks are largely unknown. In one case report, a woman treated for recurrent *C. difficile* infection experienced rapid weight gain after receiving FMT from a healthy but overweight donor (Alang and Kelly, 2015). Although donor screening processes are constantly being refined at centers performing FMT, additional risks will likely emerge with long-term study.

Recurrent *C. difficile* Infection

The primary application of FMT is as an alternative treatment approach for those with *Clostridium difficile* infection (CDI) who have failed to improve with standard antibiotic therapy. A recent metaanalysis noted that FMT is a promising therapy for recurrent CDI, with an overall efficacy of around 90% in various studies, but the safety profile has not been captured, and physicians have yet to learn which patient groups will respond most favorably (Kassam et al., 2013). The first randomized, controlled trial found 90.9% of patients receiving FMT from a donor achieved clinical cure, compared with only 62.5% of those receiving autologous FMT; no FMT-related significant adverse events were reported (Kelly et al., 2016). A detailed characterization of the gut microbial communities of those who participated in this trial found that complete donor engraftment was not necessary for clinical improvement, as long as functionally critical taxa are already present (Staley et al., 2016). To further complicate the matter, a preliminary investigation of those with CDI who, instead of FMT, received a transfer of only the sterile filtrates from donor stool, found that this filtrate (containing bacterial components, metabolites, and/or bacteriophages) was sufficient for symptom improvement (Ott et al., 2017).

Inflammatory Bowel Disease

Two recent randomized, controlled trials have added to the body of evidence on FMT for IBD—in particular, UC. FMT administered weekly for 6 weeks induced remission in those with active UC to a greater extent than placebo (water enema), with no difference in adverse events. Better efficacy was associated with the stool from one particular donor in the study and also with recently diagnosed UC patients (Moayyedi et al., 2015). But a study on individuals with UC by Rossen et al. (2015) found no significant difference in remission between those who received donor FMT and those who received autologous FMT; responders, however, were found to have distinct gut microbiota features. Several differences between the two trials may have accounted for the different results: for example, different routes of administration (enema vs nasoduodenal tube). While FMT holds significant promise for UC and is currently being investigated in ongoing studies, at present, insufficient data exists to support the routine use of FMT in those with IBD.

Other Conditions

FMT has been investigated in several other gut microbiota-linked conditions, including metabolic syndrome. In one human trial, the intestinal microbiota from lean donors was transferred to individuals with metabolic syndrome. The treatment increased both insulin sensitivity and levels of butyrate-producing bacteria, but these effects were not sustained (Vrieze et al., 2012). Given the data showing that intestinal microbiota diversity predicts mortality in allogeneic hematopoietic stem cell transplantation (allo-HSCT) recipients (Taur et al., 2014), ongoing clinical trials are investigating FMT following allo-HSCT for enhancing clinical outcomes and increasing survival.

DEFINED MICROBIAL CONSORTIA

Several companies are advancing alternatives to FMT: defined microbial consortia to treat recurrent CDI and other conditions. These could comprise a number of bacteria harvested from the gut and would be classified as probiotic treatments (Hill et al., 2014) but tested and regulated as drugs. An initially publicized commercial attempt to use a "synthetic" FMT for treating recurrent CDI fell below expectations (van der Lelie et al., 2017). Although such efforts to date have been disappointing, this line of development will continue because of the unknown long-term risks of FMT as well as the need for standardization and safety, insights into mode of action, and eventual requirements by regulatory agencies.

"Designer" probiotic formulations that have the broader aim of modulating the human immune response are currently under development by several companies, but so far, only a subset of patients appear to respond to such therapies. This problem draws attention to the enormous complexity of the gut microbiome and the need for better understanding the interactions between key strains so as to improve engraftment in the intestines for a therapeutic effect (van der Lelie et al., 2017).

Microbial consortia may be developed in various ways: for example, by isolating pure cultures of stool microorganisms from healthy donors—like those who previously donated stool to cure patients with recurrent CDI—and using them to create a defined mixture with potential therapeutic application (Martz et al., 2015). Recently, researchers from Germany used a new approach to develop a mouse gut bacterial community with enhanced colonization resistance (Brugiroux et al., 2016). They began with a community of 12 bacterial strains that stably colonized mouse intestines but conferred only partial protection against infection by the human pathogen *Salmonella enterica* serovar Typhimurium. They then performed functional genomic analysis and compared the functional potential of the 12-strain community with that of a conventional gut microbiota. From this, they identified missing functions in the 12-strain community that enabled them to create an improved version of the community that included three more facultative anaerobic bacteria. When this new community was established in germ-free mice, it successfully provided colonization resistance, showing that designing a bacterial community based on functional potential could be an effective way to pinpoint groups of bacteria that could modify health in important ways.

FURTHER METHODS OF GUT MICROBIOTA MANIPULATION FOR THERAPEUTIC BENEFIT

Gastric Bypass Surgery

Bariatric surgery is an effective option for those with a BMI of 40 or higher, but surprisingly, researchers have not fully elucidated how the procedure successfully induces long-term weight loss. An intriguing study in 2013 found that roux-en-Y gastric bypass (RYGB) restructured the intestinal microbiota in mice in a way that sham surgery did not, with a sustained increase in *Escherichia* and *Akkermansia*. Microbiota transfer from mice that had undergone RYGB was uniquely able to induce weight loss in recipient germ-free mice (Liou et al., 2013). Then, a 2015 study of women who had undergone bariatric surgery 10 years previously (and had maintained a lower weight since that time) showed persistent changes in gut microbiota compared with before the surgery: thus, a stable post-surgery weight was associated with a distinct gut microbiota composition and function (Tremaroli et al., 2015). Mechanistically, bile acids are under investigation in these effects. In mice, gastric bypass surgeries increase the circulation of bile acids that correlate with postsurgery weight loss (Myronovych et al., 2014), and the beneficial metabolic effects of these surgeries seem to require intact signaling through bile acid receptor

FXR—which appears to drive some of the postsurgery changes in gut microbiota (Zhang et al., 2016a).

Traditional Chinese Foods and Medicines

Chinese metagenomics projects of the past decade have included the modern study of traditional Chinese foods and medicines that have been used therapeutically for hundreds of years. One study found, in children with obesity, a diet high in whole grains, traditional Chinese medicinal foods, and prebiotics induced significant weight loss and lowered systemic inflammation, also leading to structural and functional changes in the gut microbiota. When transferred to germ-free mice, the preintervention microbiota induced higher inflammation than the postintervention microbiota (Zhang et al., 2015).

Furthermore, animal models have shown the promise of berberine, a component of the Chinese herb *Coptis chinensis*, as a way to improve type 2 diabetes. Berberine was found to prevent the development of obesity and insulin resistance in rats on a high-fat diet; the rats showed decreased food intake, with both SCFA-producing bacteria and fecal SCFA concentrations being elevated (Zhang et al., 2012).

Helminths

The WHO estimates that 1.5 billion individuals around the world are infected with soil-transmitted helminths (Anon., 2017), and interest is growing in the immune-modulatory effects of these parasitic worms and their relationship with the gut microbiota. Helminth colonization in individuals from Malaysia was associated with greater species richness in the gut microbiota and a greater number of observed operational taxonomic units, with enrichment of Paraprevotellaceae (Ramanan et al., 2016).

Helminths have been linked with reduced prevalence of several diseases, including allergy, IBD, and celiac disease. IBD, for example, is less prevalent in helminth-endemic regions of the world. Connections between disease risk and helminth-led modulation of the gut microbiota are being explored in rodent studies—in a mouse model of Crohn's disease, helminth infection enhanced colonization resistance to an inflammatory *Bacteroides* species, protecting genetically susceptible mice from intestinal abnormalities (Ramanan et al., 2016). Researchers in another study exposed people with celiac disease to helminths—which are shown to improve gluten tolerance in these individuals by suppressing pro-inflammatory responses—and noted that the subsequent increase in microbial species richness upon gluten challenge could be part of the mechanism by which hookworms regulate gluten-induced inflammation (Giacomin et al., 2015).

Exclusive Enteral Nutrition

Gut microbiome alterations have been observed in pediatric Crohn's disease in the context of improvements induced by exclusive enteral nutrition (EEN, delivering all of the diet directly to the gastrointestinal tract, often by tube feeding) (Quince et al., 2015). In one study, researchers were able to predict sustained remission after EEN based on the microbial communities at baseline; those who did not experience sustained remission had a notably large Proteobacteria component to their microbial communities (Dunn et al., 2016).

Probiotics in Health

Benefits of probiotic intake are difficult to show in healthy individuals, except through well-designed, long-term studies. However, in addition to the potential therapeutic uses of probiotics as detailed in this chapter, some evidence exists for the use of probiotics to prevent various aspects of disease. Detailed below are the conditions for which preventative use of probiotics may be beneficial.

Upper Respiratory Tract Infections

Upper respiratory tract infections (URTIs), including the common cold, are most often caused by viruses. A Cochrane systematic review found an advantage of probiotics over placebo for preventing URTIs and for reducing antibiotic prescription rates for acute URTIs. Side effects included minor gastrointestinal symptoms (Hao et al., 2011).

Antibiotic-Associated Diarrhea

Antibiotic-associated diarrhea (AAD) is the frequent watery bowel movements and abdominal pain that may occur in conjunction with antibiotic ingestion. In children between the ages of 0 and 18 treated with antibiotics, there appeared to be a benefit of probiotics in preventing AAD. In particular, *Lactobacillus rhamnosus* or *Saccharomyces boulardii* at a dose of 5–40 billion colony forming units/day may be appropriate (Goldenberg et al., 2015).

Allergic Conditions and Eczema

A recent systematic review and metaanalysis showed that probiotics reduced eczema in infants when used by their mothers in the third trimester of pregnancy or while breastfeeding; probiotics did not appear to affect the incidence of allergies (Cuello-Garcia et al., 2015).

REFERENCES

Akbari, V., Hendijani, F., 2016. Effects of probiotic supplementation in patients with type 2 diabetes: systematic review and meta-analysis. Nutr. Rev. 74 (12), 774–784. Available from: http://www.ncbi.nlm.nih.gov/pubmed/27864537.

Alang, N., Kelly, C.R., 2015. Weight gain after fecal microbiota transplantation. Open Forum Infect. Dis. 2 (1), ofv004. Available from: https://academic.oup.com/ofid/article-lookup/doi/10.1093/ofid/ofv004.

AlFaleh, K., et al., 2011. Probiotics for prevention of necrotizing enterocolitis in preterm infants. In: AlFaleh, K. (Ed.), Cochrane Database of Systematic Reviews, John Wiley & Sons, Chichester p. CD005496. Available from: http://www.ncbi.nlm.nih.gov/pubmed/21412889.

Allen, S.J., et al., 2003. Probiotics for treating infectious diarrhoea. In: Allen, S.J. (Ed.), Cochrane Database of Systematic Reviews, John Wiley & Sons, Chichester p. CD003048. Available from: http://www.ncbi.nlm.nih.gov/pubmed/15106189.

Anon, 2017. WHO | Soil-transmitted helminth infections. WHO. Available from: http://www.who.int/mediacentre/factsheets/fs366/en/.

Azcarate-Peril, M.A., et al., 2017. Impact of short-chain galactooligosaccharides on the gut microbiome of lactose-intolerant individuals. Proc. Natl. Acad. Sci. U. S. A. 114 (3), E367–E375. Available from: http://www.ncbi.nlm.nih.gov/pubmed/28049818.

Bass, N.M., et al., 2010. Rifaximin treatment in hepatic encephalopathy. N. Engl. J. Med. 362 (12), 1071–1081. Available from: http://www.nejm.org/doi/abs/10.1056/NEJMoa0907893.

Bernaola Aponte, G., et al., 2010. Probiotics for treating persistent diarrhoea in children. In: Bernaola Aponte, G. (Ed.), Cochrane Database of Systematic Reviews, John Wiley & Sons, Chichester p. CD007401. Available from: http://www.ncbi.nlm.nih.gov/pubmed/21069693.

Brugiroux, S., et al., 2016. Genome-guided design of a defined mouse microbiota that confers colonization resistance against *Salmonella enterica* serovar Typhimurium. Nat. Microbiol. 2, 16215. Available from: http://www.nature.com/articles/nmicrobiol2016215.

Cani, P.D., Van Hul, M., 2015. Novel opportunities for next-generation probiotics targeting metabolic syndrome. Curr. Opin. Biotechnol. 32, 21–27. Available from: http://linkinghub.elsevier.com/retrieve/pii/S0958166914001748.

Chiu, W.-C., et al., 2015. Synbiotics reduce ethanol-induced hepatic steatosis and inflammation by improving intestinal permeability and microbiota in rats. Food Funct. 6 (5), 1692–1700. Available from: http://xlink.rsc.org/?DOI=C5FO00104H.

Cho, Y.A., Kim, J., 2015. Effect of probiotics on blood lipid concentrations: a meta-analysis of randomized controlled trials. Medicine 94 (43), e1714. Available from: http://www.ncbi.nlm.nih.gov/pubmed/26512560.

Chumpitazi, B.P., et al., 2015. Randomised clinical trial: gut microbiome biomarkers are associated with clinical response to a low FODMAP diet in children with the irritable bowel syndrome. Aliment. Pharmacol. Ther. 42 (4), 418–427. Available from: http://doi.wiley.com/10.1111/apt.13286.

Coccorullo, P., et al., 2010. Lactobacillus reuteri (DSM 17938) in infants with functional chronic constipation: a double-blind, randomized, placebo-controlled study. J. Pediatr. 157 (4), 598–602. Available from: http://www.ncbi.nlm.nih.gov/pubmed/20542295.

Costalos, C., et al., 2008. The effect of a prebiotic supplemented formula on growth and stool microbiology of term infants. Early Hum. Dev. 84 (1), 45–49. Available from: http://www.ncbi.nlm.nih.gov/pubmed/17433577.

Coudray, C., et al., 1997. Effect of soluble or partly soluble dietary fibres supplementation on absorption and balance of calcium, magnesium, iron and zinc in healthy young men. Eur. J. Clin. Nutr. 51 (6), 375–380. Available from: http://www.ncbi.nlm.nih.gov/pubmed/9192195.

Cuello-Garcia, C.A., et al., 2015. Probiotics for the prevention of allergy: a systematic review and meta-analysis of randomized controlled trials. J. Allergy Clin. Immunol. 136 (4), 952–961. Available from: http://www.ncbi.nlm.nih.gov/pubmed/26044853.

Daillère, R., et al., 2016. *Enterococcus hirae* and *Barnesiella intestinihominis* facilitate cyclophosphamide-induced therapeutic immunomodulatory effects. Immunity 45 (4), 931–943. Available from: http://www.ncbi.nlm.nih.gov/pubmed/27717798.

Dinan, T.G., et al., 2013. Psychobiotics: a novel class of psychotropic. Biol. Psychiatry 74 (10), 720–726. Available from: http://www.ncbi.nlm.nih.gov/pubmed/23759244.

Dinleyici, E.C., Vandenplas, Y., 2014. *Lactobacillus reuteri* DSM 17938 effectively reduces the duration of acute diarrhoea in hospitalised children. Acta Paediatr. 103 (7). Available from: http://doi.wiley.com/10.1111/apa.12617.

Dunn, K.A., et al., 2016. Early changes in microbial community structure are associated with sustained remission after nutritional treatment of pediatric Crohn's disease. Inflamm. Bowel Dis. 22 (12), 2853–2862. Available from: http://www.ncbi.nlm.nih.gov/pubmed/27805918.

Elahi, B., et al., 2008. On the benefit of probiotics in the management of pouchitis in patients underwent ileal pouch anal anastomosis: a meta-analysis of controlled clinical trials. Dig. Dis. Sci. 53 (5), 1278–1284. Available from: http://www.ncbi.nlm.nih.gov/pubmed/17940902.

Eswaran, S.L., et al., 2016. A randomized controlled trial comparing the low FODMAP diet vs. modified NICE guidelines in US adults with IBS-D. Am. J. Gastroenterol. 111 (12), 1824–1832. Available from: http://www.nature.com/doifinder/10.1038/ajg.2016.434.

FAO/WHO, 2001. Health and nutritional properties of probiotics in food including powder milk with live lactic acid bacteria. Available from: ftp://ftp.fao.org/es/esn/food/probio_report_en.pdf.

Forslund, K., et al., 2015. Disentangling type 2 diabetes and metformin treatment signatures in the human gut microbiota. Nature 528 (7581), 262–266. Available from: http://www.nature.com/doifinder/10.1038/nature15766.

Garofoli, F., et al., 2014. The early administration of *Lactobacillus reuteri* DSM 17938 controls regurgitation episodes in full-term breastfed infants. Int. J. Food Sci. Nutr. 65 (5), 646–648. Available from: http://www.tandfonline.com/doi/full/10.3109/09637486.2014.898251.

Giacomin, P., et al., 2015. Experimental hookworm infection and escalating gluten challenges are associated with increased microbial richness in celiac subjects. Sci. Rep. 5 (1), 13797. Available from: http://www.ncbi.nlm.nih.gov/pubmed/26381211.

Gibson, G.R., et al., 2017. Expert consensus document: The International Scientific Association for Probiotics and Prebiotics (ISAPP) consensus statement on the definition and scope of prebiotics. Nat. Rev. Gastroenterol. Hepatol.

Gibson, G.R., Roberfroid, M.B., 1995. Dietary modulation of the human colonic microbiota: introducing the concept of prebiotics. J. Nutr. 125 (6), 1401–1412. Available from: http://www.ncbi.nlm.nih.gov/pubmed/7782892.

Gibson, G.R., et al., 2010. Dietary prebiotics: current status and new definition. Available from: http://centaur.reading.ac.uk/17730/.

Goldenberg, J.Z., et al., 2013. Probiotics for the prevention of *Clostridium difficile*-associated diarrhea in adults and children. In: Johnston, B.C. (Ed.), Cochrane Database of Systematic Reviews, John Wiley & Sons, Chichester p. CD006095. Available from: http://www.ncbi.nlm.nih.gov/pubmed/23728658.

Goldenberg, J.Z., et al., 2015. Probiotics for the prevention of pediatric antibiotic-associated diarrhea. In: Johnston, B.C. (Ed.), Cochrane Database of Systematic Reviews. John Wiley & Sons, Chichester. Available from: http://doi.wiley.com/10.1002/14651858.CD004827.pub4.

Halmos, E.P., et al., 2015. Diets that differ in their FODMAP content alter the colonic luminal microenvironment. Gut 64 (1), 93–100.

Hao, Q., et al., 2011. Probiotics for preventing acute upper respiratory tract infections. In: Dong, B.R. (Ed.), Cochrane Database of Systematic Reviews, John Wiley & Sons, Chichester p. CD006895. Available from: http://www.ncbi.nlm.nih.gov/pubmed/21901706.

Harb, T., et al., 2016. Infant colic—what works. J. Pediatr. Gastroenterol. Nutr. 62 (5), 668–686. Available from: http://www.ncbi.nlm.nih.gov/pubmed/26655941.

Hill, C., et al., 2014. Expert consensus document: The International Scientific Association for Probiotics and Prebiotics consensus statement on the scope and appropriate use of the term probiotic. Nat. Rev. Gastroenterol. Hepatol. 11 (8), 506–514. Available from: http://www.nature.com/doifinder/10.1038/nrgastro.2014.66.

Huang, R., Wang, K., Hu, J., 2016. Effect of probiotics on depression: a systematic review and meta-analysis of randomized controlled trials. Nutrients 8 (8). Available from: http://www.ncbi.nlm.nih.gov/pubmed/27509521.

Ianiro, G., Tilg, H., Gasbarrini, A., 2016. Antibiotics as deep modulators of gut microbiota: between good and evil. Gut 65 (11), 1906–1915. Available from: http://gut.bmj.com/lookup/doi/10.1136/gutjnl-2016-312297.

Kadlec, R., et al., 2014. The effect of prebiotics on adherence of probiotics. J. Dairy Sci. 97 (4), 1983–1990. Available from: http://www.ncbi.nlm.nih.gov/pubmed/24485681.

Kasińska, M.A., Drzewoski, J., 2015. Effectiveness of probiotics in type 2 diabetes: a meta-analysis. Pol. Arch. Med. Wewn. 125 (11), 803–813. Available from: http://www.ncbi.nlm.nih.gov/pubmed/26431318.

Kassam, Z., et al., 2013. Fecal microbiota transplantation for *Clostridium difficile* infection: systematic review and meta-analysis. Am. J. Gastroenterol. 108 (4), 500–508. Available from: http://www.ncbi.nlm.nih.gov/pubmed/23511459.

Kelly, C.R., et al., 2016. Effect of fecal microbiota transplantation on recurrence in multiply recurrent *Clostridium difficile* infection. Ann. Intern. Med. 165 (9), 609. Available from: http://annals.org/article.aspx?doi=10.7326/M16-0271.

Khalesi, S., et al., 2014. Effect of probiotics on blood pressure: a systematic review and meta-analysis of randomized, controlled trials. Hypertension 64 (4), 897–903. Available from: http://www.ncbi.nlm.nih.gov/pubmed/25047574.

Khan, K.J., et al., 2011. Antibiotic therapy in inflammatory bowel disease: a systematic review and meta-analysis. Am. J. Gastroenterol. 106 (4), 661–673. Available from: http://www.ncbi.nlm.nih.gov/pubmed/21407187.

Kristensen, N.B., et al., 2016. Alterations in fecal microbiota composition by probiotic supplementation in healthy adults: a systematic review of randomized controlled trials. Genome Med. 8 (1), 52. Available from: http://genomemedicine.biomedcentral.com/articles/10.1186/s13073-016-0300-5.

Liou, A.P., et al., 2013. Conserved shifts in the gut microbiota due to gastric bypass reduce host weight and adiposity. Sci. Transl. Med. 5 (178), 178ra41. Available from: http://www.ncbi.nlm.nih.gov/pubmed/23536013.

Macfarlane, S., Macfarlane, G.T., Cummings, J.H., 2006. Review article: prebiotics in the gastrointestinal tract. Aliment. Pharmacol. Ther. 24 (5), 701–714. Available from: http://doi.wiley.com/10.1111/j.1365-2036.2006.03042.x.

Marteau, P., 2001. Safety aspects of probiotic products. Näringsforskning 45 (1), 22–24. Available from: https://www.tandfonline.com/doi/full/10.3402/fnr.v45i0.1785.

Martz, S.-L.E., et al., 2015. Administration of defined microbiota is protective in a murine Salmonella infection model. Sci. Rep. 5, 16094. Available from: http://www.ncbi.nlm.nih.gov/pubmed/26531327.

Miquel, S., et al., 2013. *Faecalibacterium prausnitzii* and human intestinal health. Curr. Opin. Microbiol. 16 (3), 255–261. Available from: http://linkinghub.elsevier.com/retrieve/pii/S1369527413000775.

Moayyedi, P., et al., 2015. Fecal microbiota transplantation induces remission in patients with active ulcerative colitis in a randomized controlled trial. Gastroenterology 149 (1), 102–109.e6. Available from: http://linkinghub.elsevier.com/retrieve/pii/S0016508515004515.

Myronovych, A., et al., 2014. Vertical sleeve gastrectomy reduces hepatic steatosis while increasing serum bile acids in a weight-loss-independent manner. Obesity 22 (2), 390–400. Available from: http://www.ncbi.nlm.nih.gov/pubmed/23804416.

Nikfar, S., Darvish-Da, M., Abdollahi, M., 2010. A review and meta-analysis of the efficacy of antibiotics and probiotics in management of pouchitis. Int. J. Pharmacol. 6 (6), 826–835. Available from: http://www.scialert.net/abstract/?doi=ijp.2010.826.835.

Nordic Council of Ministers, 2000. Safety Evaluation of Fructans. Nordic Council of Ministers, Copenhagen.

Ohta, A., et al., 1995. Effects of fructooligosaccharides on the absorption of iron, calcium and magnesium in iron-deficient anemic rats. J. Nutr. Sci. Vitaminol. 41 (3), 281–291. Available from: http://www.ncbi.nlm.nih.gov/pubmed/7472673.

Ojetti, V., et al., 2014. The effect of *Lactobacillus reuteri* supplementation in adults with chronic functional constipation: a randomized, double-blind, placebo-controlled trial. J. Gastrointestin. Liver Dis. 23 (4), 387–391. Available from: http://www.ncbi.nlm.nih.gov/pubmed/25531996.

Ott, S.J., et al., 2017. Efficacy of sterile fecal filtrate transfer for treating patients with *Clostridium difficile* infection. Gastroenterology 152 (4), 799–811.e7. Available from: http://linkinghub.elsevier.com/retrieve/pii/S0016508516353549.

Park, S., Bae, J.-H., 2015. Probiotics for weight loss: a systematic review and meta-analysis. Nutr. Res. 35 (7), 566–575. Available from: http://www.ncbi.nlm.nih.gov/pubmed/26032481.

Pimentel, M., et al., 2011. Rifaximin therapy for patients with irritable bowel syndrome without constipation. N. Engl. J. Med. 364 (1), 22–32. Available from: http://www.nejm.org/doi/abs/10.1056/NEJMoa1004409.

Pirbaglou, M., et al., 2016. Probiotic supplementation can positively affect anxiety and depressive symptoms: a systematic review of randomized controlled trials. Nutr. Res. 36 (9), 889–898. Available from: http://www.ncbi.nlm.nih.gov/pubmed/27632908.

Plovier, H., et al., 2016. A purified membrane protein from *Akkermansia muciniphila* or the pasteurized bacterium improves metabolism in obese and diabetic mice. Nat. Med. 23 (1), 107–113. Available from: http://www.nature.com/doifinder/10.1038/nm.4236.

Quince, C., et al., 2015. Extensive modulation of the fecal metagenome in children with Crohn's disease during exclusive enteral nutrition. Am. J. Gastroenterol. 110 (12), 1718–1729. Available from: http://www.ncbi.nlm.nih.gov/pubmed/26526081.

Rahimi, R., et al., 2006. A meta-analysis of broad-spectrum antibiotic therapy in patients with active Crohn's disease. Clin. Ther. 28 (12), 1983–1988. Available from: http://www.ncbi.nlm.nih.gov/pubmed/17296455.

Rahimi, R., et al., 2008. A meta-analysis on the efficacy of probiotics for maintenance of remission and prevention of clinical and endoscopic relapse in Crohn's disease. Dig. Dis. Sci. 53 (9), 2524–2531. Available from: http://www.ncbi.nlm.nih.gov/pubmed/18270836.

Ramanan, D., et al., 2016. Helminth infection promotes colonization resistance via type 2 immunity. Science 352 (6285), 608–612. Available from: http://www.ncbi.nlm.nih.gov/pubmed/27080105.

Ritchie, M.L., et al., 2012. A meta-analysis of probiotic efficacy for gastrointestinal diseases. In: Heimesaat, M.M. (Ed.), PLoS One 7 (4), e34938. Available from: http://dx.plos.org/10.1371/journal.pone.0034938.

Rossen, N.G., et al., 2015. Findings from a randomized controlled trial of fecal transplantation for patients with ulcerative colitis. Gastroenterology 149 (1), 110–118.e4. Available from: http://www.ncbi.nlm.nih.gov/pubmed/25836986.

Sarkar, A., et al., 2016. Psychobiotics and the manipulation of bacteria-gut-brain signals. Trends Neurosci. 39 (11), 763–781. Available from: http://www.ncbi.nlm.nih.gov/pubmed/27793434.

Sartor, R.B., et al., 2004. Therapeutic manipulation of the enteric microflora in inflammatory bowel diseases: antibiotics, probiotics, and prebiotics. Gastroenterology 126 (6), 1620–1633. Available from: http://linkinghub.elsevier.com/retrieve/pii/S0016508504004561.

Schreck Bird, A., et al., 2016. Probiotics for the treatment of infantile colic: a systematic review. J. Pharm. Pract. Available from: http://www.ncbi.nlm.nih.gov/pubmed/26940647.

Schrezenmeir, J., de Vrese, M., 2001. Probiotics, prebiotics, and synbiotics—approaching a definition. Am. J. Clin. Nutr. 73 (2 Suppl.), 361S–364S. Available from: http://www.ncbi.nlm.nih.gov/pubmed/11157342.

Scott, K.P., et al., 2013. The influence of diet on the gut microbiota. Pharmacol. Res. 69 (1), 52–60. Available from: http://www.ncbi.nlm.nih.gov/pubmed/23147033.

Sharma, B.C., et al., 2013. A randomized, double-blind, controlled trial comparing rifaximin plus lactulose with lactulose alone in treatment of overt hepatic encephalopathy. Am. J. Gastroenterol. 108 (9), 1458–1463. Available from: http://www.ncbi.nlm.nih.gov/pubmed/23877348.

Shen, J., Zuo, Z.-X., Mao, A.-P., 2014. Effect of probiotics on inducing remission and maintaining therapy in ulcerative colitis, Crohn's disease, and pouchitis: meta-analysis of randomized controlled trials. Inflamm. Bowel Dis. 20 (1), 21–35. Available from: http://www.ncbi.nlm.nih.gov/pubmed/24280877.

Shin, N.-R., et al., 2014. An increase in the *Akkermansia* spp. population induced by metformin treatment improves glucose homeostasis in diet-induced obese mice. Gut 63 (5), 727–735. Available from: http://gut.bmj.com/lookup/doi/10.1136/gutjnl-2012-303839.

Sierra, C., et al., 2015. Prebiotic effect during the first year of life in healthy infants fed formula containing GOS as the only prebiotic: a multicentre, randomised, double-blind and placebo-controlled trial. Eur. J. Nutr. 54 (1), 89–99. Available from: http://www.ncbi.nlm.nih.gov/pubmed/24671237.

Silk, D.B.A., et al., 2009. Clinical trial: the effects of a trans-galactooligosaccharide prebiotic on faecal microbiota and symptoms in irritable bowel syndrome. Aliment. Pharmacol. Ther. 29 (5), 508–518. Available from: http://www.ncbi.nlm.nih.gov/pubmed/19053980.

Staley, C., et al., 2016. Complete microbiota engraftment is not essential for recovery from recurrent *Clostridium difficile* infection following fecal microbiota transplantation. MBio 7 (6), e01965–16. Available from: http://www.ncbi.nlm.nih.gov/pubmed/27999162.

Sun, J., Buys, N., 2015. Effects of probiotics consumption on lowering lipids and CVD risk factors: a systematic review and meta-analysis of randomized controlled trials. Ann. Med. 47 (6), 430–440. Available from: http://www.ncbi.nlm.nih.gov/pubmed/26340330.

Taur, Y., et al., 2014. The effects of intestinal tract bacterial diversity on mortality following allogeneic hematopoietic stem cell transplantation. Blood 124 (7), 1174–1182. Available from: http://www.ncbi.nlm.nih.gov/pubmed/24939656.

Teuri, U., Korpela, R., 1998. Galacto-oligosaccharides relieve constipation in elderly people. Ann. Nutr. Metab. 42 (6), 319–327. Available from: http://www.ncbi.nlm.nih.gov/pubmed/9895419.

Torres, D.P.M., et al., 2010. Galacto-oligosaccharides: production, properties, applications, and significance as prebiotics. Compr. Rev. Food Sci. Food Saf. 9 (5), 438–454. Available from: http://doi.wiley.com/10.1111/j.1541-4337.2010.00119.x.

Tremaroli, V., et al., 2015. Roux-en-Y gastric bypass and vertical banded gastroplasty induce long-term changes on the human gut microbiome contributing to fat mass regulation. Cell Metab. 22 (2), 228–238. Available from: http://www.ncbi.nlm.nih.gov/pubmed/26244932.

Tursi, A., et al., 2010. Treatment of relapsing mild-to-moderate ulcerative colitis with the probiotic VSL#3 as adjunctive to a standard pharmaceutical treatment: a double-blind,

randomized, placebo-controlled study. Am. J. Gastroenterol. 105 (10), 2218–2227. Available from: http://www.ncbi.nlm.nih.gov/pubmed/20517305.
van der Lelie, D., et al., 2017. The microbiome as a source of new enterprises and job creation: considering clinical faecal and synthetic microbiome transplants and therapeutic regulation. Microb. Biotechnol. 10 (1), 4–5. Available from: http://doi.wiley.com/10.1111/1751-7915.12597.
Verna, E.C., Lucak, S., 2010. Use of probiotics in gastrointestinal disorders: what to recommend? Ther. Adv. Gastroenterol. 3 (5), 307–319. Available from: http://www.ncbi.nlm.nih.gov/pubmed/21180611.
Vetizou, M., et al., 2015. Anticancer immunotherapy by CTLA-4 blockade relies on the gut microbiota. Science 350 (6264), 1079–1084. Available from: http://www.sciencemag.org/cgi/doi/10.1126/science.aad1329.
Viaud, S., et al., 2013. The intestinal microbiota modulates the anticancer immune effects of cyclophosphamide. Science 342 (6161), 971–976. Available from: http://www.ncbi.nlm.nih.gov/pubmed/24264990.
Vrieze, A., et al., 2012. Transfer of intestinal microbiota from lean donors increases insulin sensitivity in individuals with metabolic syndrome. Gastroenterology 143 (4), 913–916.e7. Available from: http://www.ncbi.nlm.nih.gov/pubmed/22728514.
Zhang, X., et al., 2012. Structural changes of gut microbiota during berberine-mediated prevention of obesity and insulin resistance in high-fat diet-fed rats. In: Sanz, Y. (Ed.), PLoS One 7 (8), e42529. Available from: http://dx.plos.org/10.1371/journal.pone.0042529.
Zhang, C., et al., 2015. Dietary modulation of gut microbiota contributes to alleviation of both genetic and simple obesity in children. EBioMedicine 2 (8), 968–984.
Zhang, L., et al., 2016a. Farnesoid X receptor signaling shapes the gut microbiota and controls hepatic lipid metabolism. mSystems 1 (5). Available from: http://msystems.asm.org/content/1/5/e00070-16.
Zhang, Y., et al., 2016b. Effects of probiotic type, dose and treatment duration on irritable bowel syndrome diagnosed by Rome III criteria: a meta-analysis. BMC Gastroenterol. 16 (1), 62. Available from: http://www.ncbi.nlm.nih.gov/pubmed/27296254.

CHAPTER 8
Practical Diet Recommendations

Objectives

- To learn the answers to common questions pertaining to the gut microbiome that are faced by dietitians and nutritionists.
- To understand what is known about the impact of therapeutic diets on the gut microbiota.
- To gain an overview of the practical diet strategies for supporting the gut microbiome that can be incorporated into clinical practice.

Despite the plethora of websites, blogs, books, and personal testimonials claiming to have found the "holy grail" of dietary treatment for the microbiome, there are is a paucity of scientific research to support these claims. Only future research will be able to delineate the components of the optimal diet for enhancing the microbiome and health. This is not to say that eating habits do not impact the microbiome: there is scientific evidence to support the notion that lifestyle factors—in particular, the foods people eat—play a major role in shifting the microbiota, as discussed in Chapter 6. But at the present time, there is not enough evidence to make concrete dietary recommendations.

This chapter is structured around some of the most common questions related to the gut microbiome that clinicians face in their practice. Practical recommendations will be provided where possible; while it should be noted that recommendations may change and become more specific in the years ahead as scientists learn more about diet-gut microbiome interactions, Fig. 8.1 illustrates the microbiota-modulating diet strategies supported by evidence at the present time.

THERAPEUTIC DIETS AND THE INTESTINAL MICROBIOTA

Is Clinical Microbiome Testing Useful?

Commercial tests for analyzing microbiome composition are available, but are not currently useful for guiding targeted improvement of health (Davis, 2014). Since predicting current or future host phenotype from microbiota

Gut Microbiota
https://doi.org/10.1016/B978-0-12-810541-2.00008-7

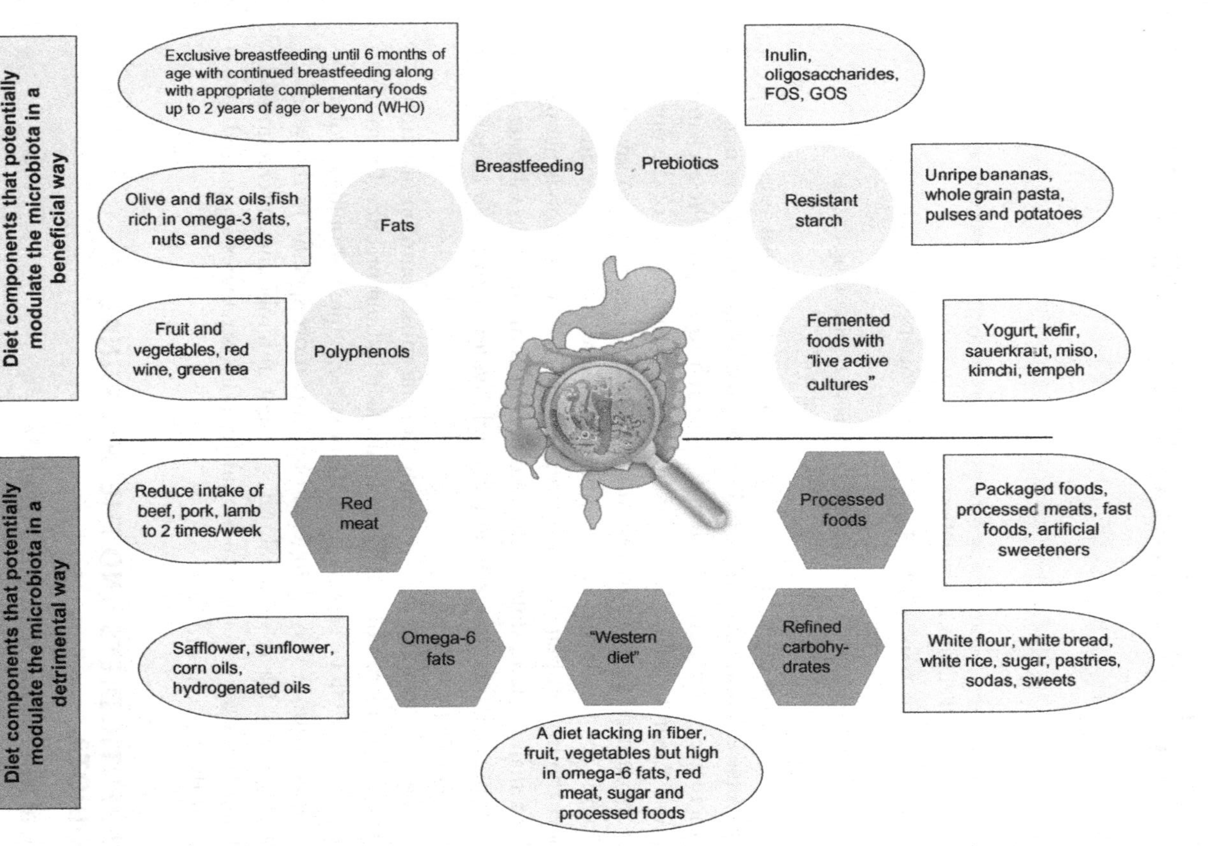

Fig. 8.1 A summary of the diet recommendations currently supported by science for maintaining a health-associated microbiota. The top half of the diagram shows diet components that may modulate the gut microbiota in a way that is beneficial to health, while the bottom half shows components that may modulate the gut microbiota in way that is detrimental to health. These recommendations will likely evolve as more studies emerge. *FOS*, fructooligosaccharides; *GOS*, galactooligosaccharides; *WHO*, World Health Organization.

composition alone is impossible given current knowledge, and since scientists know little about how to reliably bring about a stable shift in gut microbiota composition with any dietary intervention, it is unclear what information these microbiome analysis tests add to current medical and dietary assessments. On the other hand, individuals who test their microbiomes and fill out accompanying health and lifestyle questionnaires may be contributing to a "citizen science"-derived pool of gut microbiome data that could yield valuable insights about health and disease (Topol and Richman, 2016).

Is "Leaky Gut" a Medically Recognized Diagnosis That Makes a Therapeutic Diet Necessary?

An ailment purportedly involving defective tight junctions in the intestine and entry of bacteria and toxins into the bloodstream with widespread physiological effects, "leaky gut syndrome" is not a medical condition. It is true that intestinal permeability is increased in some diseases such as inflammatory bowel diseases (Gerova et al., 2011) and celiac disease (Heyman et al., 2012), but there is scant evidence to support leaky gut as a cause of these diseases or any other disease. Furthermore, there is no evidence that restrictive diets can ameliorate intestinal permeability in a way that makes a difference to health.

Does Short-Term Dieting Impact the Gut Microbiome?

Although it is established that weight loss is best achieved through long-term lifestyle shifts including diet and exercise (Hassan et al., 2016), many people still engage in short-term dieting: temporary changes in eating patterns, with restriction of foods (e.g., gluten, sugar, fat) perceived as "bad" by some groups (Marchessault et al., 2007). Although little is known about the long-term impact of dieting in humans, evidence from mouse models shows short-term dieting may be more detrimental to weight and metabolic health than not dieting at all. In one recent study, mice who had gained weight on a high-fat diet and then normalized their weight on a regular diet were predisposed to gaining weight more rapidly the second time they switched to a high-fat diet, compared with mice that had never dieted (Thaiss et al., 2016). The mechanism was found to be related to the gut microbiota: obesity followed by short-term dieting had left mice with a gut microbial signature that predisposed them to faster weight regain upon reexposure to a high-fat diet; this was confirmed by transferring the post-dieting gut microbiota to germ-free mice. The extent to which this gut microbiota "scarring" occurs in humans as a result of short-term dieting is an interesting area for further study.

Do High-Protein, Low-Carbohydrate Diets Impact the Microbiome?

High-protein, low-carbohydrate diets (often called "low-carb" diets) are commonly used among individuals seeking weight loss (Noble and Kushner, 2006); however, they may alter microbial activity and bacterial populations in the large intestine in a detrimental way and thus impact gut health. With this type of diet, reduced intake of fermentable carbohydrates, along with high protein intake, is common. A high-protein (~138 g/day), low-carbohydrate (22 g/day) weight loss diet provided to 17 obese men resulted in a decrease in total fecal short-chain fatty acid (SCFA) concentrations, with a disproportionate reduction in amount of butyrate (Russell et al., 2011). The increased intake of dietary protein resulted in a shift toward protein fermentation in the colon, with a marked increase in fecal *N*-nitroso compounds (which are known carcinogens).

Consistent with previous reports, when 19 obese men were provided with a high-protein (120 g/day), low-carbohydrate (24 g/day) weight loss diet, there was a significant reduction in SCFAs, especially butyrate (Duncan et al., 2007). As carbohydrate decreased, so did *Roseburia* spp., *Eubacterium rectale* subgroup of cluster clostridial XIVa, and *Bifidobacterium*. Brinkworth and colleagues showed that fecal concentrations of butyrate and total SCFAs, as well as counts of *Bifidobacterium*, were significantly lower when a low-carbohydrate weight loss diet (4% of total energy from carbohydrate) was consumed than when a high-carbohydrate diet (46% of total energy from carbohydrate) was consumed (Brinkworth et al., 2009).

A diet high in microbiota-accessible carbohydrates (MACs) may be different: Zhang and colleagues demonstrated that a diet high in fermentable carbohydrates with a balanced macronutrient profile shifted an obesity-associated dysbiotic gut microbiota to a structure with relatively lower levels of bacteria that produce potentially toxic metabolites from the fermentation of dietary fats and proteins, and with higher levels of bacteria (*Bifidobacterium* spp.) supported by the fermentation of carbohydrates. This dietary modulation of the gut microbiota contributed to the alleviation of metabolic deteriorations (Zhang et al., 2015).

These studies reveal several metabolic consequences of typical high-protein, reduced carbohydrate weight loss diets that raise concerns for gut health, especially if these diets are adopted over the long term.

Does a Gluten-Free Diet Impact the Microbiome?

Gluten is a protein component of wheat, barley, rye, and several other grains. The gluten-free diet is a medically necessary treatment for celiac disease

(CD); however, recently, the gluten-free diet has gained popularity among those in the general public attempting to lose weight, improve health, and cure the "leaky gut," without concrete evidence to support such benefits.

The impact of a gluten-free diet on the gut microbiome was studied in 10 healthy volunteers (De Palma et al., 2009). After 1 month of following the diet, there was a decrease in potentially health-promoting species of *Bifidobacterium*, as well as *Clostridium lituseburense*, *Faecalibacterium prausnitzii*, *Lactobacillus*, and *Bifidobacterium longum*, whereas *Escherichia coli*, Enterobacteriaceae, and *Bifidobacterium angulatum* were increased. De Palma and colleagues also examined immune function and found that the gluten-free diet led to a significantly lower production of TNF-α (tumor necrosis factor-alpha) and IFN-γ (interferon-gamma) and the chemokines IL (interleukin)-8 and IL-10 (De Palma et al., 2009). A decrease in the diversity of *Lactobacillus* and *Bifidobacterium* species was also reported in subjects treated with a gluten-free diet (Nistal et al., 2012). A more recent study using 16S rRNA gene sequencing in healthy subjects following a gluten-free diet for 1 month revealed that the diet induced changes in gut microbial composition (Bonder et al., 2016): decreased abundance of the family Veillonellaceae (class Clostridia) and species *Ruminococcus bromii* and *Roseburia faecis*. The families Victivallaceae, Clostridiaceae, and Coriobacteriaceae and the genus *Slackia* increased in abundance on a gluten-free diet. The changes in abundance of taxa were related to the change in diet; specifically affected were bacteria involved in carbohydrate and starch metabolism. Contrary to previous reports, however, the gluten-free diet was not found to influence inflammatory gut markers (Bonder et al., 2016). At this time, it is not clear how gluten specifically impacts the gut microbiome, and more studies are needed to assess the impact of a gluten-free diet, not only in those with existing pathologies besides celiac disease but also in healthy individuals. There is no evidence to suggest that the gluten-free diet is a cure for the "leaky gut."

Does a Low-FODMAP Diet Impact the Microbiome?

The low-fermentable oligosaccharide, disaccharide, monosaccharide, and polyol (**low FODMAP**) diet consists of eliminating foods high in fermentable but poorly absorbed carbohydrates and polyols, usually for 6–8 weeks (Gibson and Shepherd, 2010). FODMAPs comprise fructose, lactose, fructo- and galactooligosaccharides (fructans and galactans), and polyols (sorbitol, mannitol, xylitol, and maltitol); a diet particularly low in these components is an emerging treatment for the management of irritable bowel syndrome (IBS), with studies showing good clinical efficacy (Nanayakkara et al., 2016).

Despite the improvement in individuals' digestive symptoms on a low-FODMAP diet, emerging research indicates that the diet may have a negative impact on the gut microbiome due to the reduced prebiotic content (Halmos et al., 2015; McIntosh et al., 2016). When the gut microbiota of people on a diet was compared with that of people on a typical Australian diet, the low-FODMAP diet was associated with lower absolute abundance of total bacteria, butyrate-producing bacteria, and *Bifidobacterium* spp., *Akkermansia muciniphila,* and *Ruminococcus gnavus.* Marked lower relative abundances of *Clostridium* cluster XIVa and a significantly higher abundance of *Ruminococcus* torques were also observed (Halmos et al., 2015), showing altered gut microbiota degradation of mucus and altered production of butyrate. A recent study examining the colonic microbiota in IBS demonstrated that consumption of a low-FODMAP diet resulted in higher bacterial richness, specifically Firmicutes, Clostridiales, and Actinobacteria, with increased bacterial diversity within the Actinobacteria (McIntosh et al., 2016). These results confirm findings of the previous study by Halmos et al. that the low-FODMAP diet results in changes to microbial composition, but it is not fully clear whether the changes are positive or negative.

Does the "Paleo Diet" Impact the Microbiome?

One of the most popular modern diets is the **Paleolithic diet**, otherwise known as the "paleo" or "stone age" diet. The modern-day paleo diet pattern consists of unlimited intake of vegetables, fruit, lean meats, tofu, nuts, and seeds; however, dairy products, cereal/grain products, and legumes are prohibited (Pitt, 2016). Compared with a typical high-protein, low-carbohydrate diet, the paleo diet is focused more on consuming whole foods and less on eliminating all carbohydrates. The diet is based on the "evolutionary discordance hypothesis," where the belief is that changes to the traditional nutrition and activity patterns of human hunter-gatherer ancestors have contributed to the endemic chronic diseases of the modern day (Konner and Eaton, 2010). The paleo diet has been subject to intense criticism and controversy in the medical community, due to its scientifically unsubstantiated claims about weight loss and improvement of health. Although the existing scientific evidence suggests that the gut microbiome differs between those in industrialized modern societies and those in traditional hunter-gatherer societies, there is a paucity of rigorous research examining the impact of the present day paleo diet on health and on the gut microbiome. Future research is warranted, given the wide popularity of this diet.

PRACTICAL DIET RECOMMENDATIONS TO SUPPORT A HEALTH-ASSOCIATED INTESTINAL MICROBIOTA

Should Processed Foods be Reduced to Improve Gut Health?

More research examining the impact of processed foods on the gut microbiome is needed before recommendations can be provided. However, a study examining emulsifiers (carboxymethyl cellulose and polysorbate-80) in rodents found they had a detrimental influence on the gut microbiota (as shown by decreased abundance of Bacteroidales) and reduced intestinal mucosal thickness (Chassaing et al., 2015). Noncaloric artificial sweeteners (including saccharin, sucralose, and aspartame), when fed to rodents, resulted in much higher glucose intolerance and were associated with increased abundance of bacteria belonging to the genus *Bacteroides* and order Clostridiales in the gut (Suez et al., 2014). Thus, these particular components of processed foods may emerge as detrimental for the gut microbiota.

Should Fermented Foods be Consumed to Improve Gut Health?

The consumption of traditionally fermented foods, such as sauerkraut, kefir, yogurt, miso, and others has been associated with several health benefits, but direct evidence for the benefits of various foods remains limited. Intake of fermented foods has been associated with weight maintenance (Mozaffarian et al., 2011) and a reduced risk of type 2 diabetes (Chen et al., 2014) and cardiovascular disease (Tapsell, 2015), with several randomized, controlled trials supporting a causal link between fermented foods and improvement in metabolic parameters (e.g., Kim et al., 2011).

Fermented foods provide the benefits associated with nutrients contained in the foods, of course, but additional benefits may accrue from the transformation of the substrates by live microbes and/or from the presence of live microbes at the time of consumption. Various researchers have proposed that—depending on the raw materials and the microbe(s) involved in the fermentation process—a fermented food could in theory inhibit the growth of pathogens in the gut, improve food digestibility, or enhance vitamin synthesis or absorption (Marco et al., 2017). In addition, while the species of lactobacilli and bifidobacteria found in many fermented foods do not qualify as probiotics because they are uncharacterized, these species may be either identical to or share traits with known probiotic species (Marco et al., 2017); thus, a reasonable argument can be made that they provide health benefits and therefore should be consumed as part of a health-promoting diet.

Should More Fiber be Consumed to Improve Gut Health?

Some researchers have argued a lack of dietary fiber is the primary feature of a typical Western diet that has led to depletion of the gut microbiome and an increase in chronic disease, by substantially diminishing microbiota diversity and decreasing production of beneficial metabolites (Deehan and Walter, 2016). Fermentable fiber sources not only provide a substrate for the growth of microbes but also increase concentrations of bacterial fermentation products, such as SCFAs (described in previous chapters), which are necessary for both gut health and overall health. The most well-studied fermentable fiber sources to date are oligosaccharides and resistant starch.

Evidence supports the benefits of dietary fiber to such an extent that some scientists are calling for a refiguring of fiber intake recommendations in order to optimize the prevention of chronic disease (Deehan and Walter, 2016). Even at present, dietary fiber intakes in many populations are far below recommended levels. Functional foods that incorporate prebiotics may emerge as a tool for enriching the diet with health-promoting fiber.

Should Resistant Starch be Consumed to Improve Gut Health?

Resistant starch is a form of dietary fiber that resists digestion in the small intestine and reaches the colon, where it is metabolized by the microbiota, resulting in SCFA production. At this time, no specific intake recommendations can be made for resistant starch; however, a study in 1996 suggested 20 g/day were needed to confer benefits on gut health (Baghurst et al., 1996). The best sources of resistant starch in the diet include unripe bananas, pasta, pulses, and potatoes; the wholegrain versions of products such as pasta and rice are higher in resistant starch than the refined versions. Another form of resistant starch is retrograde starch, which is formed when starchy foods (e.g., potatoes or pasta) are cooked and then cooled.

Should Fruits and Vegetables be Increased to Improve Gut Health?

Nutrition professionals have been promoting the health benefits of fruits and vegetables for many years; however, now, improved gut health can be added to the rationale for promoting increased consumption of these items. Fruit and vegetables, rich in both polyphenols and fiber, likely assist in modulating the gut microbiota toward a more health-promoting profile by increasing lactobacilli and bifidobacteria.

What Types of Fats Should be Consumed for Optimal Gut Health?

There is a paucity of human research on the impact of various fats on the gut microbiota, mainly because it is difficult to study fat as an isolated component of the diet. Research in rodent models suggests olive oil and flaxseed/fish oil result in the most diverse intestinal microbiota. This is an area that will continue to evolve as more clinical data are generated.

Does Protein Derived From Animals Have a Particular Impact on Gut Health?

Changes to the intestinal microbiota have been documented with the type of dietary protein consumed. *Bacteroides* are highly associated with animal proteins and variety of amino acids, whereas *Prevotella* is highly associated with increased intakes of plant protein (Wu et al., 2011). Interventional studies have demonstrated that high-protein diets result in reductions to fecal butyrate concentrations and butyrate-producing bacteria, which have negative impacts on gut health. Although further research is needed, reducing protein from animal sources, with a concomitant increase in vegetable protein sources, may be prudent.

What Is the Best Use of Probiotics?

Scientists are still deciphering the characteristics of the gut microbiome that support optimal health. This makes it extremely challenging to provide general recommendations for healthy individuals on specific probiotic strains or products to consume. Scientists also do not yet know the optimal dosage required to provide health benefits.

A variety of probiotic strains have been studied for their preventative and therapeutic effects in disease. Resources are available to clinicians for guiding them to probiotic products in the marketplace that contain the optimal strain(s) for a given health condition. The Clinical Guide to Probiotic Supplements (available in Canada and the United States; http://www.probioticchart.ca) is a good example. When a clinician is recommending a probiotic for a specific condition, he/she should provide a strain, dosage, and product name to clients to ensure they receive maximal benefit.

How Can the Infant Microbiome be Supported?

Although many factors in early life affect the microbiota—like mode of delivery, environmental exposures, and antibiotics—diet is one of the key influencers. Breast milk offers many benefits to the human infant beyond the microbiota, but one key benefit to the microbiota is its large content of human milk oligosaccharides that specifically feed bifidobacteria. Due to the complexity of breast milk, its beneficial impacts on the gut microbiome are not fully understood; however, the scientific evidence does support the notion that breastfeeding provides the developing infant with what it needs for a health-supportive gut microbiome later in life.

Are New Clinical Nutrition Practice Guidelines Warranted?

At this point, research on dietary impacts on the gut microbiota is still in its infancy; therefore, the creation of clinical practice guidelines poses a difficulty. In the research, more emphasis needs to be placed on whole-diet intervention studies versus individual foods and their components, in order to understand how the relationships between diet, the gut microbiota, and health work in the context of a whole diet. Certainly, there is enough research to indicate that a "Western" diet with high consumption of fast foods or processed foods, high levels of omega-6 polyunsaturated fatty acids, and high sugar intake—with low fiber—is associated with negative changes to the microbiome and health. A diet rich in fiber, mainly from fruit and vegetables, low in red meat, and rich in omega-3 polyunsaturated fatty acids increases health-associated microbiota characteristics, with more Bacteroidetes and *Bifidobacterium*.

REFERENCES

Baghurst, P.A., Baghurst, K.I., Record, S.J., 1996. Dietary fibre, non-starch polysaccharides and resistant starch—a review. Food Aust. 48 (3), S3–S35.

Bonder, M.J., et al., 2016. The influence of a short-term gluten-free diet on the human gut microbiome. Genome Med. 8 (1), 45.

Brinkworth, G.D., et al., 2009. Comparative effects of very low-carbohydrate, high-fat and high-carbohydrate, low-fat weight-loss diets on bowel habit and faecal short-chain fatty acids and bacterial populations. Br. J. Nutr. 101 (10), 1493–1502.

Chassaing, B., et al., 2015. Dietary emulsifiers impact the mouse gut microbiota promoting colitis and metabolic syndrome. Nature 519 (7541), 92–96. Available from: http://www.ncbi.nlm.nih.gov/pubmed/25731162.

Chen, M., et al., 2014. Dairy consumption and risk of type 2 diabetes: 3 cohorts of US adults and an updated meta-analysis. BMC Med. 12 (1), 215.

Davis, L., 2014. Is it Really Worth Having Your Gut Bacteria Tested? Gizmodo.

De Palma, G., et al., 2009. Effects of a gluten-free diet on gut microbiota and immune function in healthy adult human subjects. Br. J. Nutr. 102 (8), 1154–1160.

Deehan, E.C., Walter, J., 2016. The fiber gap and the disappearing gut microbiome: implications for human nutrition. Trends Endocrinol. Metab. 27 (5), 239–242.

Duncan, S.H., et al., 2007. Reduced dietary intake of carbohydrates by obese subjects results in decreased concentrations of butyrate and butyrate-producing bacteria in feces. Appl. Environ. Microbiol. 73 (4), 1073–1078.

Gerova, V.A., et al., 2011. Increased intestinal permeability in inflammatory bowel diseases assessed by iohexol test. World J. Gastroenterol. 17 (17), 2211–2215.
Gibson, P.R., Shepherd, S.J., 2010. Evidence-based dietary management of functional gastrointestinal symptoms: The FODMAP approach. J. Gastroenterol. Hepatol. 25 (2), 252–258.
Halmos, E.P., et al., 2015. Diets that differ in their FODMAP content alter the colonic luminal microenvironment. Gut 64 (1), 93–100. Available from: http://www.ncbi.nlm.nih.gov/pubmed/25016597.
Hassan, Y., et al., 2016. Lifestyle interventions for weight loss in adults with severe obesity: a systematic review. Clin. Obes. 6 (6), 395–403.
Heyman, M., et al., 2012. Intestinal permeability in coeliac disease: insight into mechanisms and relevance to pathogenesis. Gut 61 (9), 1355–1364.
Kim, E.K., et al., 2011. Fermented kimchi reduces body weight and improves metabolic parameters in overweight and obese patients. Nutr. Res. 31 (6), 436–443.
Konner, M., Eaton, S.B., 2010. Paleolithic nutrition: twenty-five years later. Nutr. Clin. Pract. 25 (6), 594–602. Available from: http://ncp.sagepub.com/cgi/doi/10.1177/0884533610385702.
Marchessault, G., et al., 2007. Canadian dietitians' understanding of non-dieting approaches in weight management. Can. J. Diet. Pract. Res. 68 (2), 67–72.
Marco, M.L., et al., 2017. Health benefits of fermented foods: microbiota and beyond. Curr. Opin. Biotechnol. 44, 94–102.
McIntosh, K., et al., 2016. FODMAPs alter symptoms and the metabolome of patients with IBS: a randomised controlled trial. Gut 66, 1241–1251.
Mozaffarian, D., et al., 2011. Changes in diet and lifestyle and long-term weight gain in women and men. N. Engl. J. Med. 364 (25), 2392–2404.
Nanayakkara, W.S., et al., 2016. Efficacy of the low FODMAP diet for treating irritable bowel syndrome: The evidence to date. Clin. Exp. Gastroenterol. 9, 131–142.
Nistal, E., et al., 2012. Differences in faecal bacteria populations and faecal bacteria metabolism in healthy adults and celiac disease patients. Biochimie 94 (8), 1724–1729.
Noble, C.A., Kushner, R.F., 2006. An update on low-carbohydrate, high-protein diets. Curr. Opin. Gastroenterol. 22 (2), 153–159. Available from: http://www.ncbi.nlm.nih.gov/pubmed/16462172.
Pitt, C.E., 2016. Cutting through the Paleo hype: the evidence for the Palaeolithic diet. Aust. Fam. Physician 45 (1), 35–38.
Russell, W.R., et al., 2011. High-protein, reduced-carbohydrate weight-loss diets promote metabolite profiles likely to be detrimental to colonic health. Am. J. Clin. Nutr. 93 (5), 1062–1072.
Suez, J., et al., 2014. Artificial sweeteners induce glucose intolerance by altering the gut microbiota. Nature 514 (7521), 181–186. Available from: http://www.ncbi.nlm.nih.gov/pubmed/25231862.
Tapsell, L.C., 2015. Fermented dairy food and CVD risk. Br. J. Nutr. 113 (S2), S131–S135.
Thaiss, C.A., et al., 2016. Persistent microbiome alterations modulate the rate of post-dieting weight regain. Nature 540 (7634), 544–551.
Topol, E., Richman, J., 2016. Citizen Science and Mapping the Microbiome. Medscape.
Wu, G.D., et al., 2011. Linking long-term dietary patterns with gut microbial enterotypes. Science 334 (6052), 105–108. Available from: http://www.ncbi.nlm.nih.gov/pubmed/21885731.
Zhang, C., et al., 2015. Dietary modulation of gut microbiota contributes to alleviation of both genetic and simple obesity in children. EBioMedicine 2 (8), 968–984.

CHAPTER 9

Applications of Gut Microbiota and Nutrition Science

Objectives

- To gain perspective on the relevance of studies on the gut microbiome and nutrition to trends and directions in the food industry.
- To understand the benefits and challenges that accompany the use of microbiota-modulating ingredients (probiotics, prebiotics, and synbiotics) in foods and other products.
- To become acquainted with how beneficial microorganisms are changing ideas of food processing and food safety.

Scientific work on the gut microbiome and nutrition is of relevance to several identifiable trends and directions in the food industry. Many factors play into whether microbiota-modulating ingredients are deliberately included in foods or consumed as supplements. Live microorganisms used in food processing, as well as probiotics, prebiotics, synbiotics, and food safety are discussed below.

Food is for everyone, for daily health and prevention of future disease, not just for those with a specific clinical disorder. Yet, the development and introduction of food products with enhanced nutritional value and tangible health benefits for consumers is of great interest to those in the food industry (Tufarelli and Laudadio, 2016). Some of these products are referred to as **functional foods**. For Gibson and Williams, a functional food can be defined as a food that "is satisfactorily demonstrated to affect beneficially one or more target functions in the body beyond adequate nutrition, in a way that improves health and well-being or reduces the risk of disease" (Gibson and Williams, 2000; based on Roberfroid, 2002). In some cases, the nutritional components conferring health benefits (e.g., prebiotics) may be present in greater concentrations than would occur in nature, or they may be present in food items that would not normally contain them. Regulations applicable to these items vary greatly in different parts of the world, with a detailed discussion of these regulations falling beyond the scope of this text. But generally, in cases where a manufacturer chooses to make a health claim

Gut Microbiota
https://doi.org/10.1016/B978-0-12-810541-2.00009-9

on the label of a functional food, it may overlap in a regulatory sense with drugs that have medicinal label claims. As the range of these products grows, the line separating food and medicine is becoming less and less clear.

Dietary supplements in the United States are a regulatory category separate from conventional foods and drugs: they are not intended to prevent or cure specific diseases, nor are they intended as a replacement for food. Manufacturers may make structure/function claims about dietary supplements (describing the role of an item in affecting the normal structure or function of the human body) with scientific substantiation, and they must report serious adverse events associated with consumption. Under Canada's regulatory system, a distinction exists between a food and a natural health product (NHP) that makes a therapeutic claim; foods are regulated by the Food Directorate of Health Canada while NHPs are regulated under the Natural and Non-prescription Health Products Directorate. While a health claim can be made for a food with adequate evidence, all Canadian NHPs require a product license with clear scientific evidence to support safety and efficacy. Product license holders must also monitor and report serious adverse reactions to the product. In both countries, overstating the health benefits of a food item remains a significant risk for companies, with words listed on packaging such as "clinically proven" and "scientifically proven" requiring appropriate substantiation.

EMPHASIS ON MICROBES IN FOOD PROCESSING

Those in the food industry know consumers have always chosen to buy foods primarily based on value and taste (Zink, 1997), with the latter being paramount, so in a competitive marketplace, constant innovation based on new tastes is a necessity. From cheese, chocolate, and coffee to beer and kombucha, microorganisms are increasingly being employed as a way to modify taste sensations and yield a variety of unique foods and beverages. Even minor alterations in the microorganisms used and in the processing conditions can produce dramatically different products (Marco et al., 2017).

Fermented foods are usually defined as "foods or beverages made through controlled microbial growth and enzymatic conversions of major and minor food components" (Marco et al., 2017). The substrates can range from meat and fish to dairy, vegetables, soy beans, cereals, and fruits. Lactic acid bacteria and yeasts are the primary microorganisms that ferment raw materials containing high levels of monosaccharides and disaccharides or starch, with molds and *Bacillus* being responsible for secondary fermentation. Table 9.1 shows the microorganisms that carry out fermentation in a variety of common food items.

Table 9.1 Production of different fermented foods requires different sources of microorganisms

Food	Source of organisms	Organisms
Yogurt	Starter culture	*St. thermophilus*, *L. delbrueckii* ssp. *bulgaricus*
Cheese, sour cream	Starter culture, backslopping	*Lc. lactis*, *Lu. mesenteroides*
Sausage	Backslopping, starter culture, spontaneous	*L. sake*, *L. plantarum*, *S. carnosus*, *S. xylosus*, *P. acidilactici*
Wine	Spontaneous, starter culture	*Sa. cerevisiae*, *O. oeni*
Beer	Backslopping, starter culture	*Sa. cerevisiae* (*L. brevis*)
Bread	Starter culture	*Sa. cerevisiae*
Sourdough bread	Backslopping	*L. sanfranciscensis*, *C. humilis*
Sauerkraut or kimchi	Spontaneous	*Lu. mesenteroides*, *L. plantarum*, *L. brevis*
Olives	Spontaneous	*L. plantarum*
Soy sauce, miso	Starter culture, spontaneous	*A. soyae*, *Z. rouxii*, *T. halophilus*
Tempeh	Starter culture, backslopping	*R. oligosporus*
Natto	Starter culture, backslopping	*B. subtilis* var. *natto*

Many species contribute to the character of the final product, but the dominant groups of microorganisms are listed in the table for each food item. **Backslopping** supplies microorganisms to a new batch of fermented food through the transfer of a small amount from a previous batch of fermented food.

St., *Streptococcus*; *L.*, *Lactobacillus*; *Lc.*, *Lactococcus*; *Lu.*, *Leuconostoc*; *S.*, *Staphylococcus*; *P.*, *Pediococcus*; *Sa.*, *Saccharomyces*; *O.*, *Oenococcus*; *C.*, *Candida*; *A.*, *Aspergillus*; *Z.*, *Zygosaccharomyces*; *T.*, *Tetragenococcus*; *R.*, *Rhizopus*; *B.*, *Bacillus*.

From Marco, M.L., Heeney, D., Binda, S., Cifelli, C.J., Cotter, P.D., Foligné, B., Gänzle, M., Kort, R., Pasin, G., Pihlanto, A., Smid, E.J., Hutkins, R., 2017. Health benefits of fermented foods: microbiota and beyond. Curr. Opin. Biotechnol. 44, 94–102. Copyright 2017, with permission from Elsevier.

Interestingly, fermented foods may be working against modern consumers' negative perceptions of processed foods (Reynolds and Kenward, 2016). Processing of a food item by microbes (e.g., rather than a machine or chemical) is seen as more positive: one industry report identified fermented foods as a top product trend of 2016, citing its association with "processing the natural way." At least one US food science and technology program reported seeing a recent increase in the number of students pursuing undergraduate degree programs, with over 75% embarking on degrees related to fermentation. Interest is reportedly driven by the perceived "greater authenticity" of fermented food products (Despain, 2014).

In some fermented foods and beverages, like chocolate, sourdough bread, and coffee, live microorganisms are employed in processing but are no longer present in the final product. Others, however, contain notable quantities of live microbes when consumed. In the fermented foods that contain live cultures at the time of consumption, many of the bacterial species either are identical to species that qualify as probiotics or share physiological traits with them (Marco et al., 2017). These bacteria do not meet the definition of probiotics, but they may have "probiotic-like" properties. This could serve to increase consumer perceptions that fermented foods improve health.

Good quality management is critical to the production of commercial fermented foods; minor variations in parameters like temperature during the manufacturing process could change the strain of bacteria that dominates in fermentation, greatly affecting product consistency (Despain, 2014). And while a minimum guarantee of viable bacteria in the final product is not promised for most fermented foods, special handling is usually required to make sure they reach the consumer in the desired condition.

PROBIOTICS

As awareness grows about the need to support the gut microbiome in health and disease, so does consumer demand for probiotics. According to multiple analyses, commercial growth of probiotic foods and supplements continues at a rapid pace. Consumer understanding of probiotics' links to digestive health has dramatically heightened in the past decade (Despain, 2014), with better educated consumers increasingly wanting to learn about the most appropriate bacterial strains and doses for consumption.

New products are continually emerging to fill this demand. Besides the refrigerated probiotics sold as foods (e.g., yogurt and kefir) or supplements, shelf-stable probiotics now exist, allowing certain probiotic strains to be added to a variety of foods and beverages, from bread and chocolate to juice.

Several challenges must be addressed when incorporating probiotics into a food item. Survivability of the microorganisms is an important issue, since the majority of available probiotics are not shelf-stable; a minimum guarantee of colony-forming units present at the time of consumption may be required by regulators. Manufacturers may originally need to add 10–100 times the number of microorganisms that are needed at the time of consumption (Zink, 1997), increasing the cost of production. Additional costs may be incurred to ensure the product is handled in a way that ensures it is in the desired condition when it reaches the consumer (Sanders, 2000).

Stability of bacterial strain features in the final product must also be considered. A study in 1983 alerted manufacturers to the idea that lot-to-lot variations in preparations of *Lactobacillus acidophilus* may affect clinical outcomes (Clements et al., 1983). Since then, others have found that production and manufacturing methods have the potential to influence relevant properties of a given probiotic strain, rendering this an important consideration for quality control (Grześkowiak et al., 2011).

Safety is a critical issue as well, with probiotics requiring particular attention to safe manufacturing (Baldi and Arora, 2015). Probiotics are often subject to different standards for manufacturing and quality control than pharmaceuticals, and although they appear to be very safe in healthy individuals, some have called for careful consideration of their manufacture when there is a possibility of use in "at risk" populations. Understandable concern arose about a 2014 case of fatal gastrointestinal mucormycosis in a 3-pound preterm infant, which was reported by the Centers for Disease Control and Prevention (CDC, 2015). The infant had received a probiotic dietary supplement for the prevention of necrotizing enterocolitis—an indication for probiotics that has support in the literature. The opportunistic pathogen mold likely originated as an unintentional contaminant in the probiotic manufacturing process; this case highlighted the importance of knowing that microbiological contamination during manufacturing of probiotics may pose a risk.

PREBIOTICS

As described in Chapter 7, the scope and appropriate use of the term prebiotic is only recently a matter of scientific consensus (Gibson et al., 2017). Now that a useful and scientifically based definition of a prebiotic has been proposed, regulatory agencies in different parts of the world may move to establish or refine their requirements for the use of these ingredients in commercially available foods.

While some well-studied prebiotics such as fructooligosaccharides (FOS), galactooligosaccharides (GOS), and lactulose have a history of safe commercial use (Macfarlane et al., 2006), consumer acceptance of prebiotic-fortified products is a known challenge. Manufacturers aim to produce items that contain enough prebiotics to have the desired health impact while remaining acceptable and palatable. Research showed inulin-fortified bread, for example, resulted in a smaller loaves, a harder crumb, and a darker color than control loaves. Consumer acceptability decreased in loaves with higher inulin content, but fortification of around 5% appeared feasible (Morris and Morris, 2012). Interestingly, one study found that a prebiotic (FOS) could be added to a peach-flavored drinkable yogurt without compromising consumer acceptance, but including both a probiotic and a prebiotic (*L. acidophilus* and FOS) had a negative impact on acceptance (Gonzalez et al., 2011).

SYNBIOTICS

Synbiotics, which include at least one prebiotic and one probiotic ingredient, are also incorporated into some food items. In theory, the use of synbiotics is a strategy to enhance the establishment and competitiveness of probiotic bacteria in the digestive tract, and in practice, the combination allows a label claim of beneficial physiological effect relevant to the prebiotic (which may have stronger scientific substantiation in some cases) for a product that also happens to contain a probiotic.

One challenge in the use of synbiotics is deciding on the most appropriate pairings of probiotic and prebiotic ingredients. Currently, common available synbiotic combinations include forms of bifidobacteria and lactobacilli with FOS or inulin; a single-genus probiotic used in a synbiotic combination could consist of bifidobacteria or *Lactobacillus rhamnosus* GG with inulin, although some commercially available synbiotic products contain prebiotics that do not feed the accompanying probiotics. A recent study reported a new way for identifying probiotic strains to use in synbiotic applications: researchers administered to healthy volunteers increasing doses of a prebiotic (GOS) and then isolated bacteria from their fecal samples. They found an eightfold enrichment in *Bifidobacterium adolescentis* strain IVS-1, highlighting it as a strain with possible synergistic effects when used alongside GOS (Krumbeck et al., 2015).

NEW TOOLS FOR FOOD SAFETY

Food safety and avoidance of contamination (as discussed in Chapter 4) is of perennial importance to the food industry. The need for safer food has prompted companies to explore new food preservation and safety systems:

Table 9.2 Some microorganisms present in fermented foods produce antibacterial metabolites capable of inhibiting pathogens: these can be either bacteriocins (proteins active against closely related bacterial strains), isolates, or organic acids (carbon-containing compounds with acidic properties)

Organism(s)	Type	Secondary metabolite	Pathogen inhibited	Reference
Lactobacillus curvatus 54M16	Bacteriocin	Sakacins	*Listeria monocytogenes*, *Bacillus cereus*	Casaburi et al. (2016)
L. crispatus	Isolates	I2-31, C3-12, F-l, F-50, F-59	*Salmonella*	Kim et al. (2014)
L. paracasei	Organic acid	Citric acid (pH 2.2), glutamic acid (pH 4.2)	*Fusarium culmorum*	Zalan et al. (2009)
L. plantarum 1MAU 10124	Organic acid	Phenyl lactic acid	*Penicillium roqueforti*	Zhang et al. (2014)
L. plantarum CECT-221	Organic acid	Phenyl pyruvic acid	*S. aureus*, *R. aeruginosa*, *L. monocytogenes*, *S. enterica*	Rodriguez-Pazo et al. (2013)
Bifidobacterium longum, *B. breve*	Organic acid	Lactic acid	*Salmonella typhimurium*, *S. aureus*, *E. coli*, *E. faecalis*, *C. difficile*	Tejero-Sarinena et al. (2012)
B. infantis	Organic acid	Acetic acid	N/A	Tejero-Sarinena et al. (2012)
L. plantarum	Bacteriocin	Plantaricin	Varied serotypes of *L. monocytogenes*	Barbosa et al. (2016)
L. plantarum	Organic acid	Formic acid (pH 2.3)	*Fusarium culmorum*	Zalan et al. (2009)
L. rhamnosus	Organic acid	Succinic acid (pH 2.7)	N/A	Zalan et al. (2009)
L. lactis CL1	Bacteriocin	Pediocin	*L. monocytogenes*, *S. aureus*	Rodriguez et al. (2005)
L. lactis ESI 515	Bacteriocin	Nisin	*S. aureus*	Rodriguez et al. (2005)
L. lactis subsp. *lactis* WX153	Bacteriocin	WX153	*Streptococcus suis*	Srimark and Khunajakr (2015)

LAB, lactic acid bacteria; *N/A*, not applicable.
From Josephs-Spaulding, J., Beeler, E., Singh, O.V., 2016. Human microbiome versus food-borne pathogens: friend or foe. Appl. Microbiol. Biotechnol. 100 (11), 4845–4863,

one of these is "competitive microbial inhibition," whereby harmless or beneficial bacteria are utilized to inhibit the growth of both spoilage contaminants and pathogens (Zink, 1997). For example, inhibitory strains of lactic acid bacteria can be used in dairy cultures or refrigerated foods to improve safety while extending shelf life. These bacteria produce secondary metabolites, some of which have antimicrobial properties and prevent colonization by pathogenic microorganisms (see Table 9.2) (Josephs-Spaulding et al., 2016). Thus, increased knowledge of microorganisms in food processing is enabling a break from the traditional goal of eliminating all microorganisms. Instead, there is an emerging movement toward sustaining and perhaps empowering certain ones.

REFERENCES

Baldi, A., Arora, M., 2015. Regulatory categories of probiotics across the globe: a review representing existing and recommended categorization. Indian J. Med. Microbiol. 33 (5), 2. Available from: http://www.ncbi.nlm.nih.gov/pubmed/25657150.

Barbosa, M.S., et al., 2016. Characterization of a two-peptide plantaricin produced by *Lactobacillus plantarum* MBSa4 isolated from Brazilian salami. Food Control 60, 103–112.

Casaburi, A., et al., 2016. Technological properties and bacteriocins production by *Lactobacillus curvatus* 54M16 and its use as starter culture for fermented sausage manufacture. Food Control 59, 31–45.

Centers for Disease Control and Prevention, 2015. Notes from the field: fatal gastrointestinal mucormycosis in a premature infant associated with a contaminated dietary supplement—Connecticut, 2014. Available from: https://www.cdc.gov/mmwr/preview/mmwrhtml/mm6406a6.htm.

Clements, M.L., et al., 1983. Exogenous lactobacilli fed to man—their fate and ability to prevent diarrheal disease. Prog. Food Nutr. Sci. 7 (3–4), 29–37. Available from: http://www.ncbi.nlm.nih.gov/pubmed/6657981.

Despain, D., 2014. The new fermented food culture. Food Technol. 68 (9), 39–45.

Gibson, G.R., et al., 2017. Expert consensus document: The International Scientific Association for Probiotics and Prebiotics (ISAPP) consensus statement on the definition and scope of prebiotics. Nat. Rev. Gastroenterol. Hepatol.

Gibson, G.R., Williams, C.M., 2000. Functional Foods: Concept to Product, first ed. Woodhead, Cambridge. Available from: http://lorestan.itvhe.ac.ir/_fars/Documents/2000-22037.pdf.

Gonzalez, N.J., Adhikari, K., Sancho-Madriz, M.F., 2011. Sensory characteristics of peach-flavored yogurt drinks containing prebiotics and synbiotics. LWT Food Sci. Technol. 44 (1), 158–163. Available from: http://linkinghub.elsevier.com/retrieve/pii/S0023643810002252.

Grześkowiak, Ł., et al., 2011. Manufacturing process influences properties of probiotic bacteria. Br. J. Nutr. 105 (6), 887–894. Available from: http://www.journals.cambridge.org/abstract_S0007114510004496.

Josephs-Spaulding, J., Beeler, E., Singh, O.V., 2016. Human microbiome versus food-borne pathogens: friend or foe. Appl. Microbiol. Biotechnol. 100 (11), 4845–4863. Available from: http://link.springer.com/10.1007/s00253-016-7523-7.

Kim, J.Y., et al., 2014. Inhibition of *Salmonella* by bacteriocin-producing lactic acid bacteria derived from U.S. kimchi and broiler chicken. J. Food Saf. 35, 1–12.

Krumbeck, J.A., et al., 2015. In vivo selection to identify bacterial strains with enhanced ecological performance in synbiotic applications. Appl. Environ. Microbiol. 81 (7), 2455–2465. Available from: http://www.ncbi.nlm.nih.gov/pubmed/25616794.

Macfarlane, S., Macfarlane, G.T., Cummings, J.H., 2006. Review article: prebiotics in the gastrointestinal tract. Aliment. Pharmacol. Ther. 24 (5), 701–714. Available from: http://doi.wiley.com/10.1111/j.1365-2036.2006.03042.x.

Marco, M.L., et al., 2017. Health benefits of fermented foods: microbiota and beyond. Curr. Opin. Biotechnol. 44, 94–102. Available from: http://linkinghub.elsevier.com/retrieve/pii/S095816691630266X.

Morris, C., Morris, G.A., 2012. The effect of inulin and fructo-oligosaccharide supplementation on the textural, rheological and sensory properties of bread and their role in weight management: a review. Food Chem. 133 (2), 237–248. Available from: http://linkinghub.elsevier.com/retrieve/pii/S030881461200060X.

Reynolds, J., Kenward, E., 2016. Special report fermented foods set to flourish in 2016. Food ingredients first. Available from: http://www.foodingredientsfirst.com/news/SPECIAL-REPORT-Fermented-Foods-Set-to-Flourish-in-2016?frompage=news.

Roberfroid, M., 2002. Functional food concept and its application to prebiotics. Dig. Liver Dis. 34 (Suppl. 2), S105–S110. Available from: http://www.ncbi.nlm.nih.gov/pubmed/12408452.

Rodriguez, E., et al., 2005. Antimicrobial activity of pediocin-producing *Lactococcus lactis* on *Listeria monocytogenes, Staphylococcus aureus* and *Escherichia coli* O157:H7 in cheese. Int. Dairy J. 15, 51–57.

Rodriguez-Pazo, N., et al., 2013. Cell-free supernatants obtained from fermentation of cheese whey hydrolyzates and phenylpyruvic acid by *Lactobacillus plantarum* as a source of antimicrobial compounds, bacteriocins, and natural aromas. Appl. Biochem. Biotechnol. 171, 1042–1060.

Sanders, M.E., 2000. Considerations for use of probiotic bacteria to modulate human health 1. J. Nutr. 130, 384–390. Available from: http://www.kalbemed.com/Portals/6/kome-lib/gastrointestinal_and_hepatobiliary system/Probiotik/Synbio/considerations for use of probiotic bacteria.pdf.

Srimark, N., Khunajakr, N., 2015. Characterization of the bacteriocin-like substance from *Lactococcus lactis* subsp. *lactis* WX153 against swine pathogen *Streptococcus suis*. J Health Res. 29, 259–267.

Tejero-Sarinena, S., et al., 2012. In vitro evaluation of the antimicrobial activity of a range of probiotics against pathogens: evidence for the effects of organic acids. Anaerobe 18, 530–538.

Tufarelli, V., Laudadio, V., 2016. J. Exp. Biol. Agric. Sci. Available from: https://www.cabdirect.org/cabdirect/abstract/20163217866.

Zalan, Z., et al., 2009. Production of organic acids by *Lactobacillus* strains in three different media. Eur. Food Res. Technol. 230, 395–404.

Zhang, X., et al., 2014. A new high phenyl lactic acid-yielding *Lactobacillus plantarum* IMAU10124 and a comparative analysis of lactate dehydrogenase gene. FEMS Microbiol. Lett. 356, 89–96.

Zink, D., 1997. The impact of consumer demands and trends on food processing. Emerg. Infect. Dis. 3 (4). Available from: https://wwwnc.cdc.gov/eid/article/3/4/pdfs/97-0408.pdf.

CHAPTER 10

The Future of Gut Microbiota and Nutrition

Objectives

- To understand the current knowledge gaps and future directions of gut microbiota research—both in general and in specific relationship to nutrition.
- To foresee how gut microbiota research may change nutrition practice.

Despite the significant amount of data gathered through large-scale gut microbiota research projects and the millions of dollars spent to date, many questions remain. It is believed that numerous actionable insights about the gut microbiome and nutrition still lie ahead. Besides the current applications of probiotics, prebiotics, and fecal microbiota transplantation, clinicians and industry still struggle to provide the public with tangible products and recommendations. Surely, however, the next 5–10 years of gut microbiota research will see clinicians and companies developing new tools for addressing health.

Microbiome-based therapeutics to address specific disease states and symptoms are under development by companies around the world, while functional foods incorporating probiotics or prebiotics, for consumption by healthy individuals, are of great interest to food manufacturers. Both types of products will be necessary to optimize health for the greatest number of individuals based on new gut microbiota research.

In some contexts, scientists have characterized the microorganisms colonizing the human digestive tract as a distinct but invisible "organ" (Brown and Hazen, 2015). As understanding of the gut microbiota grows among the general public, people may indeed come to see it as an organ—or they may eventually understand it through another metaphorical framework. Clinicians will play a role in this evolving understanding, one patient conversation at a time.

Gut Microbiota
https://doi.org/10.1016/B978-0-12-810541-2.00010-5

THE FUTURE OF GUT MICROBIOME RESEARCH

At present, it is difficult to find a major chronic disease that has *not* been linked in some way to the gut microbiota. The many preliminary connections are leading to both justified excitement and, on occasion, risky overstatement. It is clear that gut microbiome science will evolve in the years ahead to yield an improved understanding of how microbes influence overall health, but likely, disturbances in gut microbiota structure and function will play a causal role in only a few of the ailments that are currently associated with gut microbiota dysbiosis. This will ultimately provide further insights about the potential—and limits—of diet to influence overall health.

Gaining Better Resolution in Gut Microbiota Studies

Throughout this text, many of the described studies have found patterns in gut bacterial composition at the phylum or genus level. Studying bacteria at this gross level may not provide sufficient insights to allow the discovery of links with disease. Thus, researchers will need to invest in methods that allow characterization of microbial communities to the level of species or even strains. Only with this increased resolution will scientists be able to start teasing apart the bacterial groups that together significantly impact disease and how these groups respond to environmental influences (Brito and Alm, 2016).

Moving Beyond Bacteria

Pipelines for characterizing the nonbacterial members of gut microbial communities are in their infancy. As this area advances, researchers will gain a better understanding of key eukaryotes, viruses/bacteriophage, and archaea that produce stability or instability in the microbial ecosystem, and the transkingdom interactions that may be relevant to health and disease (Filyk and Osborne, 2016).

Observing Real-Time Gut Microbiota Dynamics

Evidence to date shows variability in gut microbiota structure and function over time in a single environment—but it is not yet known whether these changes have implications for health and disease. Emerging tools for monitoring the gut microbiota in real time (Earle et al., 2015; Geva-Zatorsky et al., 2015; Ziegler et al., 2015) will be valuable in the future for understanding the time scales over which relevant changes in gut microbiota occur, providing details on how microbes interact with host cells and each other.

Returning to Culture-Dependent Microbiology

Metagenomics has led to the generation of many gene sequences that are not assigned to a known microorganism (Lagier et al., 2016); thus, researchers need to return to culturing specific microbes of interest in a community to gain insights into their specific properties and how they exert important effects on the community as a whole and on the host (Marx, 2016). Scientists are increasingly emphasizing the need to use information from **microbial culturomics**, an approach that uses a combination of culturing, mass spectrometry techniques, and 16S rRNA gene sequencing and has resulted in the isolation of up to double the number of species from the human gut (Lagier et al., 2016).

Focusing on Microbial Ecology

Every microbe acts within its ecological context—this is a key idea for interventions aiming to change health through the microbiota. Principles of ecology have great potential to influence how scientists understand gut microbes, yet many studies have only gone as far as cataloging compositional changes. Current probiotic therapies, for instance, are based on the idea that adding a single species or a group of similar species to an established, diverse ecosystem will have some effect on health. Limits on the effectiveness of these current therapies could be addressed, in part, through a greater understanding of gut ecology and the species interactions that may effect a change in health status. While incorporating these ideas will require advanced computational techniques, failure to consider microbial ecology will likely hinder progress in harnessing microbes to influence human health. Moreover, aspects of microbial community behavior that are specific to the gut environment (e.g., keystone species with specific activities in the GI tract) must be considered, especially when it comes to future applications of fecal microbial therapy and synthetic stool therapy (together referred to as **microbial ecosystems therapeutics** or MET) (Allen-Vercoe et al., 2012).

Aiming for a Systems Perspective

Increasingly, it is clear that gut microbiota is part of a dynamic system operating throughout the body. No longer can changes in gut microbiota composition be studied alone to yield significant insights; scientists must measure how these changes interact with other measures—such as production of specific metabolites—over space and time. Systems biology approaches use multiple platforms in order to survey global cellular processes (see Fig. 10.1). The techniques include the classic "omics" technologies such as transcriptomics, proteomics, and metabolomics, with new mathematical methods and computational tools that integrate multiple data types.

In systems biology approaches, scientists systematically characterize relationships between different components by constructing a network to aid understanding. From proteomics data, for example, they can build a protein-protein interaction network. And once these biological networks

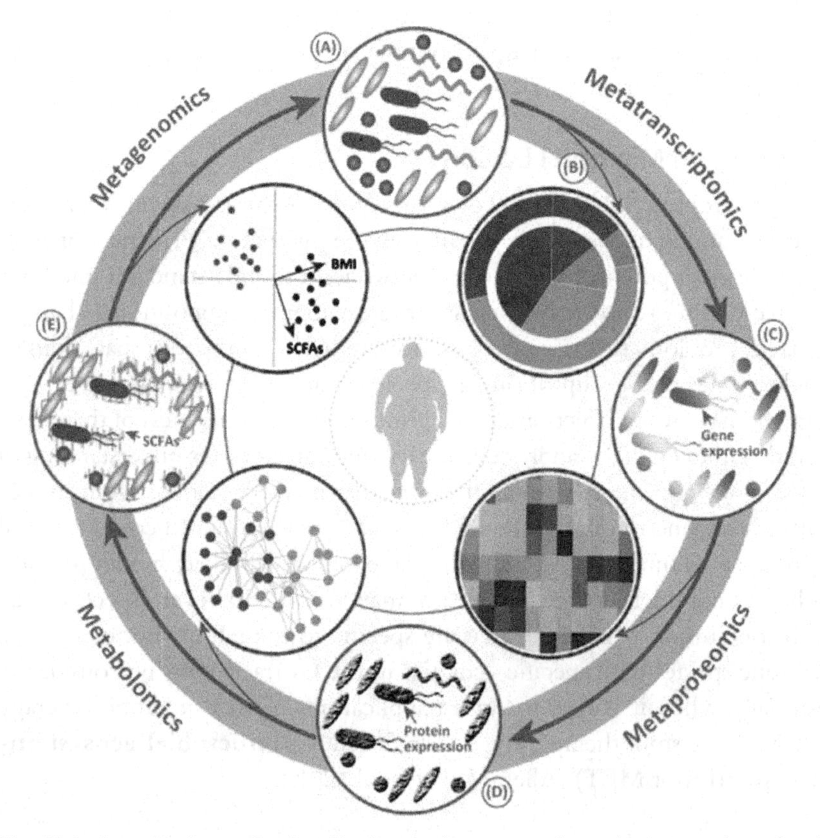

Fig. 10.1 Complex host-microbe-diet interactions necessitate the integration of multiple "omics" technologies in order to elucidate the mechanistic roles of the gut microbiota and yield insights about treating disease. Information to be integrated includes (A) taxonomic changes, (B) metagenomics data, (C) microbial gene expression (metatranscriptomics), (D) protein expression (metaproteomics), and (E) metabolites, such as short-chain fatty acids (SCFAs), linked to gut microbiota perturbations (metabolomics). *BMI*, body mass index. *(From H. Wu, V. Tremaroli, F. Bäckhed, Linking microbiota to human diseases: a systems biology perspective, Trends Endocrinol. Metab. 26 (12) (2015) 758–770, Copyright 2015, with permission from Elsevier.)*

are made, many different methods can be utilized to make this information yield valuable analyses. When it comes to the human microbiome in general, Borenstein has highlighted the urgent need for predictive system-level models (Borenstein, 2012). These approaches require a significant investment of resources but may be enabled by increased collaboration and sharing of data.

Mining the Gut Microbiome for New Drugs

Scientists currently have only a preliminary grasp of the range of bioactive chemicals secreted by the gut microbiota. A greater understanding of these microbial products is an area that holds huge potential for drug discovery: enabling a new generation of therapeutics by harnessing the "microbial pharmacists" in the gut (Spanogiannopoulos et al., 2016).

THE FUTURE OF NUTRITION

Without a doubt, the continued insights afforded by gut microbiome research will influence the way individuals think about nutrition and practice in the years ahead. Gut microbiota science will not overturn the knowledge and recommendations that have been developed from the past decades of robust research on nutrition, but the mechanistic insights it affords will be a critical part of understanding the full impact of food in the human body.

Changing Perspectives on Nutritional Assessment Relevant to Health

The relationship between certain nutritional measures and human health may change with further insights on diet and gut microbiota. Even the simple concept of increased caloric intake leading to increased adiposity, while still broadly true, is proving to depend (in some cases) on factors related to the gut microbiota (Krajmalnik-Brown et al., 2012). Thus, clinicians of the future may need to account for the role of the gut microbiota when completing a nutrition assessment on a client. The specific relevance of calories, glycemic index, and other nutritional parameters and their role in nutritional assessment will evolve with further study.

Developing Personalized Nutritional Approaches

Certain dietary approaches are effective for some individuals but not others. Available data suggest the effect of diet on host health may be personalized—and probably depends on an individual's mix of gut microbes, as determined by both genetic and environmental factors (described in Chapter 5). With the evidence necessitating a more personalized nutrition approach, general dietary recommendations may no longer be sufficient for all individuals. Techniques allowing rapid analysis of gut microbiota that inform clinicians about bacterial taxa or functional features of the microbiota could dramatically change a dietitian's practice, as he/she may be able to provide highly personalized nutrition recommendations based on this information (Harvie et al., 2016).

Leveraging Diet to Address Disease

Currently, there is little data to support the use of diet to address specific disease symptoms. But as research allows more insights about mechanisms linking gut microbiota to disease, researchers may reexamine some of the existing disease categories and diagnostic principles under the current medical model, enabling them to stratify patient populations differently and target effective treatments to each one (Wu et al., 2015). To gain insights into causality, for example, Shoaie and colleagues proposed the "community and systems-level interactive optimization" (CASINO) computational platform as a way to model the effect of diet intervention on the gut microbiota and host metabolism (Shoaie et al., 2015). Tools like this may enable scientists to better predict how altering various dietary components will affect health; diet may be among the gut-microbiota-modulating interventions recommended for subsets of those with a particular disease. Particularly in early life, dietary interventions may have an increased potential to affect lifelong health.

Using Diet to Modulate or Maintain the Gut Microbiota of Healthy Individuals

Once scientists know which gut microbiome features specifically signal the emergence of disease, it may be possible to focus on preventing disease through dietary modulation of the gut microbiota (Brahe et al., 2016). Food scientists may thus develop functional (nutritionally adapted) foods for restoring proper gut microbiota function, and these could be recommended according to the findings of regular gut microbiota monitoring. Probably, this approach will be especially important in early life, in older age, and in groups of indigenous peoples and others in urgent need of sustainable solutions for better health. Along with the growing awareness of diet's relevance to health and disease will come an increase in the importance of dietitians' involvement in aspects of medical care throughout the life span.

REFERENCES

Allen-Vercoe, E., et al., 2012. A Canadian Working Group report on fecal microbial therapy: microbial ecosystems therapeutics. Can. J. Gastroenterol. 26 (7), 457–462. Available from: http://www.ncbi.nlm.nih.gov/pubmed/22803022.

Borenstein, E., 2012. Computational systems biology and in silico modeling of the human microbiome. Brief. Bioinform. 13 (6), 769–780. Available from: http://www.ncbi.nlm.nih.gov/pubmed/22589385.

Brahe, L.K., Astrup, A., Larsen, L.H., 2016. Can we prevent obesity-related metabolic diseases by dietary modulation of the gut microbiota? Adv. Nutr. 7 (1), 90–101. Available from: http://www.ncbi.nlm.nih.gov/pubmed/26773017.

Brito, I.L., Alm, E.J., 2016. Tracking strains in the microbiome: insights from metagenomics and models. Front. Microbiol. 7, 712. Available from: http://www.ncbi.nlm.nih.gov/pubmed/27242733.

Brown, J.M., Hazen, S.L., 2015. The gut microbial endocrine organ: bacterially derived signals driving cardiometabolic diseases. Annu. Rev. Med. 66, 343–359. Available from: http://www.ncbi.nlm.nih.gov/pubmed/25587655.

Earle, K.A., et al., 2015. Quantitative imaging of gut microbiota spatial organization. Cell Host Microbe 18 (4), 478–488. Available from: http://www.ncbi.nlm.nih.gov/pubmed/26439864.

Filyk, H.A., Osborne, L.C., 2016. The multibiome: the intestinal ecosystem's influence on immune homeostasis, health, and disease. EBioMedicine 13, 46–54. Available from: http://linkinghub.elsevier.com/retrieve/pii/S2352396416304625.

Geva-Zatorsky, N., et al., 2015. In vivo imaging and tracking of host–microbiota interactions via metabolic labeling of gut anaerobic bacteria. Nat. Med. 21 (9), 1091–1100. Available from: http://www.nature.com/doifinder/10.1038/nm.3929.

Harvie, R., et al., 2016. Using the human gastrointestinal microbiome to personalize nutrition advice: are registered dietitian nutritionists ready for the opportunities and challenges? J. Acad. Nutr. Diet. 110, 48–51. Available from: http://www.ncbi.nlm.nih.gov/pubmed/27986518.

Krajmalnik-Brown, R., et al., 2012. Effects of gut microbes on nutrient absorption and energy regulation. Nutr. Clin. Pract. 27 (2), 201–214. Available from: http://www.ncbi.nlm.nih.gov/pubmed/22367888.

Lagier, J.-C., et al., 2016. Culture of previously uncultured members of the human gut microbiota by culturomics. Nat. Microbiol. 1, 16203. Available from: http://www.nature.com/articles/nmicrobiol2016203.

Marx, V., 2016. Microbiology: the return of culture. Nat. Methods 14 (1), 37–40. Available from: http://www.nature.com/doifinder/10.1038/nmeth.4107.

Shoaie, S., et al., 2015. Quantifying diet-induced metabolic changes of the human gut microbiome. Cell Metab. 22 (2), 320–331. Available from: http://www.ncbi.nlm.nih.gov/pubmed/26244934.

Spanogiannopoulos, P., et al., 2016. The microbial pharmacists within us: a metagenomic view of xenobiotic metabolism. Nat. Rev. Microbiol. 14 (5), 273–287. Available from: http://www.nature.com/doifinder/10.1038/nrmicro.2016.17.

Wu, H., Tremaroli, V., Bäckhed, F., 2015. Linking microbiota to human diseases: a systems biology perspective. Trends Endocrinol. Metab. 26 (12), 758–770. Available from: http://linkinghub.elsevier.com/retrieve/pii/S1043276015001940.

Ziegler, A., et al., 2015. Single bacteria movement tracking by online microscopy—a proof of concept study. In: Driks, A. (Ed.), PLoS One 10 (4), e0122531. Available from: http://dx.plos.org/10.1371/journal.pone.0122531.

INDEX

Note: Page numbers followed by *f* indicate figures, *t* indicate tables and *b* indicate boxes.

Printed in the United States
By Bookmasters